Biomedical Aspects of the Laser

The Introduction of Laser Applications
Into Biology and Medicine

LEON GOLDMAN, M.D.

Director, Laser Laboratories
Children's Hospital Research Foundation
and Medical Center, University of Cincinnati

Professor and Chairman
Department of Dermatology
College of Medicine
University of Cincinnati

SPRINGER SCIENCE+BUSINESS MEDIA, LLC 1967

Laser Laboratory, Children's Hospital Research Foundation
Supported by the John A. Hartford Foundation

Studies In Protection of Laser Personnel
Supported by a Grant, U.S. Public Health Service OH-00118

ISBN 978-3-540-03811-5   ISBN 978-3-642-85797-3 (eBook)
DOI 10.1007/978-3-642-85797-3

© 1967 by Springer Science+Business Media New York
Originally published by Springer-Verlag New York Inc. in 1967
Softcover reprint of the hardcover 1st edition 1967

Library of Congress Catalog Card Number 67-19308

Title No. 1427

Preface

This book is a review of past and current studies and future plans of the Laser Laboratory in Cincinnati and some of the contributions of laser research groups in other medical centers.

Special thanks are due to the Directing Physicist of the Laser Laboratory, R. James Rockwell. Without his advice, constant supervision and corrections, this enthusiastic investigator would continue to upset even many more people than he has done already. The excuse, of course, is to stimulate much needed interest and controlled research and development of the laser for biology and medicine. The Associate Research Physicist, Ralph Schooley, has worked with many phases of laser research but especially in Q spoiling, Raman spectroscopy, and the almost alchemy of holography. Holography, as of now, provides many opportunities for Gumperson's Law, "If anything can go wrong, it will."

Sincere appreciation is expressed to the Surgeons in the Laser Laboratory, who have supplied clinical and investigative surgical supervision often under great difficulties, Dr. V. E. Siler and Dr. Bruce Henderson. We are grateful for help from the Directing Biologist of the Laser Laboratory, Edmond Ritter, the Director of Laser Neurosurgery, Dr. Thomas Brown and the Professor of Neurosurgery, Dr. Robert McLaurin, for important and basic work in laser neurosurgery. Special thanks are given to Robert Meyer, who has given most of the treatments in careful and skillful fashion, and his associate, Robert Otten. In various ways, with wisecracks, chewing gum and candy, they have calmed the apprehensive adult and unwary child thrust into the gloomy atmosphere of the laser laboratory designed primarily for safety, non-specular reflections and not from calm coloring. The Surgical Technician, Daniel Crout, whose careful and precise preparation of animals has assisted in the animal research work. The engineers in the Laboratory, Robert Epstein and Domenic Carnevale, have made dream instruments alive, and the histologist, Marilyn Franzen, makes beautiful sections, especially in histochemistry. Billie Wilson, office manager of the Department of Dermatology, and her assistant, Hazel Meyer, have waded through numerous manuscripts, notes and incomprehensible hieroglyphics to provide legible copies. Kay Runck, the office manager of the Laser Laboratory and Mary McCravy, her assistant, unwound many an endless tape over many overtime hours. All these are from our own group who have worked hard in the preparation of this.

To my associates in Dermatology who are interested in and have worked with me on the laser, Dr. Donald Blaney and Dr. Karl Wm. Kitzmiller, and recently Dr. James Eha and Dr. Herman Solomon. All have contributed help, original work and efforts in this field.

From outside the Laboratory we are grateful to Dr. Walter Geeraets for his review of the very important research in the eye phases, Dr. Charles Campbell for his photography, Dr. Christian Zweng for his data, also Dr. Frederick Brech has done our laser spectroscopy, Dr. William Buckley gave generously material for laser photography, Dr. Donald Rounds for his work on cytology and cytogenetics, Dr. Josef Warkany and Dr. Shirley Soukup for their help also in cytogenetics.

A special debt is owed Dr. Edward Pratt, Professor of Pediatrics at the Medical Center of the University of Cincinnati and Director of the Children's Hospital Research Foundation in which the Laser Laboratory is located. His interest and guidance for many of the administrative headaches of the Laboratory has been greatly appreciated. He organized and guided the Laser Advisory Board of the Medical Center: Dean, Clifford G. Grulee, Dr. Wm. A. Altemeier, Professor of Surgery and Chairman of the Department. Dr. V. E. Siler, Professor of Surgery and Director of Laser Surgery, and Dr. Robt. McLaurin, Professor of Neurosurgery.

This book shows the needs and interests of many disciplines to prepare and to work for the laser of the future.

Contents

Introduction

The purpose of this book is to introduce many disciplines to the preliminary studies in many fields of biology and medicine. The laser continues to excite imagination and produce more colorful speculation than actual work especially in the various fields of medicine. In the physics laboratory, in industry and in the military, extensive studies continue. It is obvious even to the casual reader that much of the biomedical material, unlike the other fields of laser technology, is still developing in early infancy. Most of the investigations in biology and medicine are pilot studies and probings often without uncontrolled features. Yet, all these are to try to determine where one goes and what goals one sets up in attempting to investigate and to plan for the future in biomedical laser work.

As is common with any new thing, especially a new colorful material or a gadget or an instrument, be it as simple as a magnifier or as sophisticated and expensive as a laser, the initial phases of overenthusiasm swing to overpessimism. Only a few earnest researchers who have stayed on the once overcrowded, blatant, colorful, flashing band wagon continue. Some have complained with predetermined poor taste induced by anorexia nervosa or other motives, that much biomedical research is but a confused hash, or rehash, of repetitive, overwarmed, tasteless pot boilers. Yet, much of the work is new and each advance, especially with new instruments, points provokingly toward an avenue of thinking for future laser research. Since there are many differences in the many optical systems used as test models, whether they be animals of one species or another or man, there will be disagreements and controversy.

We trust that the book will point out the directions in which to proceed more reliably and safely with better instruments, better studies of biophysics, better basic studies and better controls and better techniques of laser treatment and of observation.

The book progresses from an introduction of some of the current laser instrumentation, its difficulties, its safety measures, its reliabilities, its lack of flexibility and its terrific expense, through various techniques needed to monitor this instrument and how to use it. As research and development in laser instrumentation continues, special books will be needed on this phase alone. Finally, the book comes into these areas which may be called basic research. The references suggest reading which may be done to amplify many of these very briefly reported statements. The review continues, then, through animal research and to human research and the significant phases of the laser eye program. This book will show to many scientists, such as, physicists, biologists, instrument engineers, health officers, especially those in the field of radiobiology, cytologists, cytogeneticists, physicians and surgeons, some backgrounds and what trends they can follow to investigate the use of the laser in their own field. It will be seen from all of this that as yet, there is no simple solution. The laser is not a routine or conventional instrument. Only certain medical centers can have this awful responsibility of clinical applications of the laser at present. We hope that such restrictions will not last long. Many cytological laboratories can have the laser microscope now. In the meantime, those who have the terrific responsibility to do controlled, safe and reliable work and to develop trend should not be discouraged by the frightened, the disinterested, the uninformed, the jealous and those hostile to anything new. Those in biomedical laser research, then, must follow in a critical fashion and the significant advances that the physicists and engineers have made and to attempt to adopt laser technology for the betterment of mankind.

There is still much to be done with this amazingly precise and powerful instrument.

1

Laser History and Theory

INTRODUCTION

Historically, the development of the laser concept should be considered as the natural progression of the research studies conducted in the early 1950's involving microwave amplifiers. That the lasers derive their basis from maser theory is reflected by the fact that they were for sometime commonly referred to as "optical masers." The word "LASER" is an acronym for "Light Amplification by Stimulated Emission of Radiation," and describes the emission process by which this intense beam of electromagnetic radiation is generated.

Einstein first discussed the concept of stimulated emission in 1917. During the period from 1917–1950, however, the problems of developing a satisfactory quantum theory were of sufficient magnitude that the stimulated emission phenomenon was neglected by the experimentalists in favor of the more central issues of atomic and molecular spectroscopy.

The use of stimulated emission for microwave amplification was suggested independently by Weber at the University of Maryland, by Basov and Prokhorov in the Soviet Union and by Townes at Columbia University during the period 1953–1954. Townes, Basov and Prokhorov received the Nobel Prize in Physics in 1964 for this and their subsequent work in quantum electrodynamic theory. As early as 1946, the process of stimulated emission had been observed, but not recognized. It was observed in nuclear induction experiments that an inverted resonance occurred in the special case of "adiabatic fast passage." This condition was later recognized as one of the possible methods for obtaining the population inversion required for stimulated emission.

Thus, it was not until 1958, when a paper by Schawlow and Townes was presented on the feasibility of producing stimulated emission in the microwave area near optical regions of the spectrum, that interest was renewed. Two years later, Maiman, then at Hughes Aircraft Company,

1

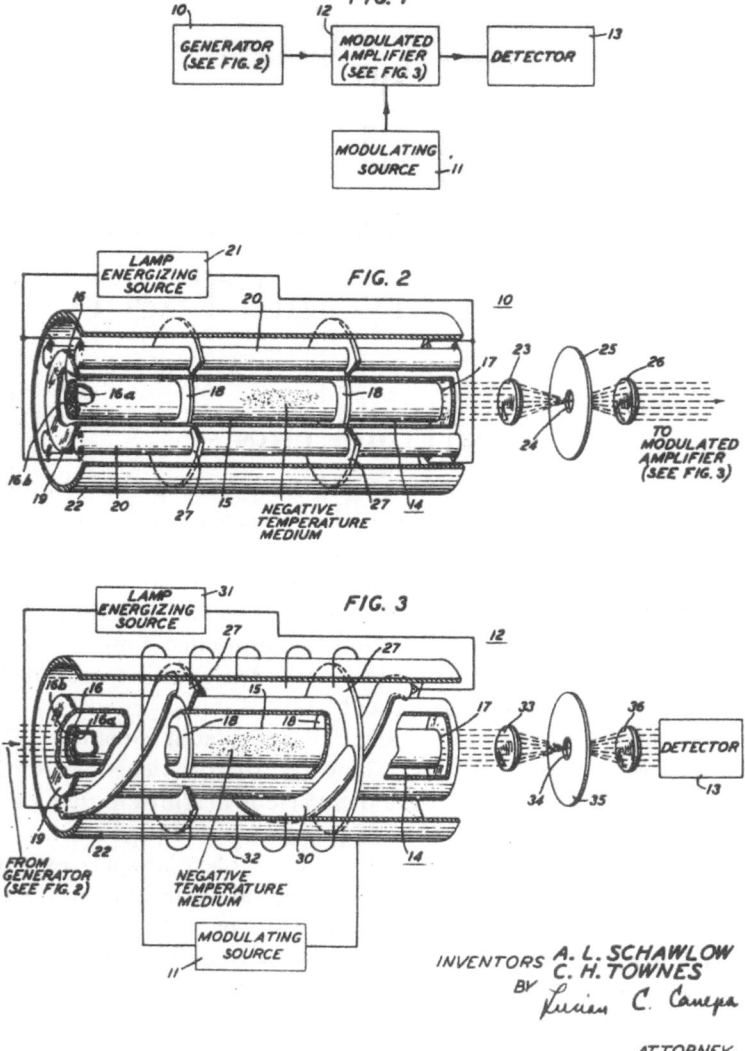

FIG. 1

FIG. 2

FIG. 3

INVENTORS A. L. SCHAWLOW
C. H. TOWNES
BY *Lucian C. Canepa*

ATTORNEY

successfully produced the first working model of the ruby laser. The search began for other suitable solid-state laser materials, but to date this has revealed only a few.

Continuous wave (CW) laser operation was first demonstrated by Javan in 1961, with the helium-neon gas laser. Since then, a multitude of gases have been investigated and utilized as the principal active laser medium. Argon, krypton, and xenon gases have been successfully operated on the ion laser principle, while carbon dioxide gas has shown the most efficient operation from a molecular collision process. Until 1965, the output of gas lasers had been very low, and, consequently, had seen little use in biomedical research. In the last two years, technological developments have permitted an increase in outputs of 5 to 10 orders of magnitude and, because of this, considerable interest has been generated in their possible medical applications.

LASER ACTION

Laser action is perhaps most easily described by relating the process which occurs in the ruby laser: the excitation of the chromium ions in the ruby is achieved by optically "pumping" the ruby rod. This process is one of the ways to achieve a change of state in an atom and involves the absorption of a photon of an exact energy value (i.e., exact wavelength). In general, atoms can occupy any one of various quantum states or energy levels depending upon the ambient conditions. If a sufficient amount of energy at the proper frequency is absorbed by the atoms of the laser media (i.e., chromium in ruby, neodymium in glass, argon, carbon dioxide, etc.) a condition called population inversion may be achieved. This term describes the condition wherein there exist more atoms in an excited energy state than are found in the ground state, and is a necessary condition for stimulated emission. In the case of the ruby laser, this relates to the acquisition and storage of more atoms in the metastable state than are in the ground state. If the material is "pumped" with enough energy, storage will occur because the average lifetime of an atom in the metastable state is longer than that of an atom in the higher-energy excited state.

The energy lost in dropping from the excited state to the metastable level is released through vibrations in the crystal lattice, and is eventually dissipated in the form of heat. The final drop to the ground state results in energy dissipation in the form of radiation at a lower frequency than that of the absorbed light. In laser devices, this process is one in which the loss occurs through the stimulated emission process or, as Schawlow indicates, "radiation of the proper frequency, corresponding to a transition between this state and some lower one, can stimulate emission of the same frequency. The excited atom thereby makes a transition to the lower state and delivers its energy to the wave."

With these conditions, light may be amplified as a result of the stimulated emission process. The light spontaneously emitted as the atom decays from the metastable state to the ground state may "stimulate" the remaining atoms in the metastable state to decay and thus emit light. The photons emitted by such emission will, in turn, stimulate further emission, and this increasing result develops in such a manner that an enormous number of photons (light quanta) are given off in a very short period of time. The duration of this rapid emission is called the laser pulse.

COHERENCE

One of the unique properties of laser radiation is the factor of coherence. As stated by Heavens, ". . . if waves from two regions of space can produce interference effects, they are described as coherent. It is generally assumed that the time of observation of the fringe pattern is long compared with the life-times of atoms from which radiation is derived." Two coherent photons can be roughly compared to two tuning forks vibrating exactly together at the same frequency and precisely in phase.

The difference between stimulated emission and spontaneous emission is that in the former the atom is caught while still in the excited state and is induced by another photon to give off radiation. This stimulated process produces a light beam with a high degree of spatial (phase) and temporal (time) coherence. Spontaneous emission occurs at random and with no spatial or temporal coherence properties of the emitted photons.

SOLID STATE LASER SYSTEMS

The basic requirements for a solid laser is that it must possess the following:

1. Reasonably broad absorption band through which energy can enter the system.
2. The crystal must fluoresce in a discrete frequency.
3. The fluorescence lifetime must be as long as possible compared with the other relevant relaxation absorption processes.
4. The crystal should be nearly perfect and free of strain defects.
5. The crystal should be reasonably durable with faces finished optically flat.
6. The crystal should be a good thermal conductor.

The solid material employed for the active laser media should exhibit reasonable sharp fluorescent lines in order that a sufficient radiant power per unit band width can be stored and enable the stimulated emission effects to dominate. This is done by doping the host material with a small percentage of atoms for which optical transitions can occur between fairly well-shielded electrons. For example, the doping in ruby of chromium atoms is typically 0.05%. The absorption spectra of such ruby displays predominate peaks in the green-violet portion of the visible spectrum (4000 Å and 5500 Å).

In the neodymium type of solid-state laser, calcium tungstate or glass may be used as the host material into which neodymium ions are doped. In this case, doping of the neodymium ions is typically 4%.

To provide the excitation energy necessary for optical pumping process, xenon flash lamps are generally used in most solid-state systems. These lamps are either straight or helically shaped quartz tubes filled with xenon gas at low pressure. Electrical energy can rapidly be discharged through such lamps from a capacitor bank charged to high voltage. To produce the resonator characteristic necessary to support oscillations in the ruby rod, multiple dielectric coatings are deposited directly on the faces of the laser crystal. Usually in small laser systems the back mirror will have a 99% reflectivity and the exit mirror will be coated for 50% to 75% reflectivity. In larger systems no coatings are used on the front face as the Fresnel reflections are sufficient to sustain oscillation.

The mirrors are so coated that the light will be reflected under optimum conditions of parallelism. The rod is placed parallel to a linear flash lamp and both are within a cylinder with the inner walls of mirror polish. The cross-section of the cylinder is often elliptical, in which case the flash lamp is placed at one focal point of the ellipse and the laser rod at the other. The flash lamp is then pulsed for a brief period, causing the crystal to emit a burst of coherent laser radiation only a few hundred microseconds after initiation of the lamp discharge.

Until 1961 the pulse duration of the solid-state laser outputs was of the order of milliseconds. However, in 1961 McClung and Hellwarth, utilizing a technique now known as Q-switching or Q-spoiling, shortened the pulse length to nanoseconds, consequently achieving enormous peak powers in the laser output.

Hercher defines Q-switching as ". . . that mode of operation of a laser in which an excess population inversion over that required for normal laser operation is attained before regenerative amplification is permitted to begin." The "Q" (quality) of the resonating cavity is reduced to prevent premature oscillation. Then, at the appropriate time, the higher Q is abruptly restored. A very high peak power output is accomplished by the result of the excess population inversion. To develop Q-switching, a rotating synchronized mirror, a Kerr cell shutter, or a saturable filter is used. The Q-switched laser emission may have a peak power in the range from 10 to 1000 megawatts, a narrow angular beam width, and a narrow spectral band width.

This technique has opened up an entirely new field of nonlinear optics and has permitted physicists to study in detail many fascinating nonlinear optical processes. From this work have evolved such phenomena as harmonic generation, parametric interactions, stimulated Raman and stimulated Brillouin scattering, two photon absorption and the so-called self-focussing effects. There is every indication that Q-switched lasers will remain an important tool in the physicists' laboratories.

Q-switched lasers have also been used, and continue to be used, as a vaporizing source in emission spectroscopy. They represent an excellent technique for easily vaporizing minute samples of biological materials for spectroscopic analysis. Ultimately, it is hoped that such a laser system will represent a unique tool for making analysis of human material in vivo.

Solid-state lasers are notoriously inefficient in their conversion of input electrical energy to the output light energy, the maximum efficiency in ruby being about 1% and in neodymium somewhat higher at 4%.

GAS LASER SYSTEMS

In this system, for the excitation for population inversion, a glow discharge is excited either by direct current flowing through the gas or by capacitatively coupled high-frequency discharge. The original gas laser, developed by Javan, utilized a low-pressure mixture of helium-neon which was excited by electron collision in a radio frequency discharge tube. Recent gas lasers with the high outputs necessary in biomedical work include the argon and carbon dioxide. Research is now being done on the nitrogen and neon lasers for production of ultraviolet wavelengths of high power outputs.

The argon laser operating at 4880 Å has been investigated for cutting applications in tissues. At power levels between 2 and 5 watts, bloodless incisions can be performed with a focussed argon laser beam in even highly vascularized structures.

SEMICONDUCTOR DIODE LASERS

The introduction of semiconductor lasers in 1962 generated considerable interest because of their considerably higher efficiencies. Since the first gallium-arsenide laser, a variety of other semiconductor lasers has been produced which can be operated in a pulsed or continuous wave manner.

The diode is a structure which allows electric current to pass in one direction through the device, but offers high electrical resistance in the reverse direction. Using gallium-arsenide upon which has been deposited a thin coating of zinc, a junction diode may be formed. This design allows the current to flow more readily in the direction from the zinc to the gallium-arsenide. The laser reaction occurs in this juction region. As mentioned previously, the output of these lasers is too low for current research in biomedical applications. Also, operation at liquid helium or liquid hydrogen temperature is generally required.

These lasers will have applications in industrial systems of the future as well as having use in electronic biomedical equipment and intracavity use in patients. It is important to emphasize, however, that in these junction diode lasers, still very much in their infancy, the radiation is generated by direct conversion of electrical energy.

CONCLUSIONS

In brief, there is now available for research in the fields of biology and medicine, a light source of highly spectral purity or monochromaticity which can be collimated to a fine degree with high radiation densities, small angular divergences and, in addition, a long coherent time. This all too brief introduction to the physics and principles of laser action shows that the biologist and physician are in constant need of the services of the laser physicist. For a detailed review of the physics of the laser, the reader is referred to the many reports and symposia on this subject. Here, in this chapter, we have given only a brief and spotted review of those points of interest in biomedical laser research.

REFERENCES

Basov, N. G. and A. M. Prokhorov: Applications of molecular beams to radio spectroscopic studies of rotation spectra of molecules. *J. Exp. Theor. Physics* (USSR), **27**:431, 1954.

Bishop, Bruce: The laser. *CoOp Engineer*, University of Cincinnati. Dec., 1965.

Bloembergen, N.: Proposal for a new type solid-state maser. *Phys. Rev.*, **104**:324, 1956.

Franken, R. A., A. E. Hill, C. W. Peters, and G. Weinreich: Generation of optical harmonics. *Phys. Rev. Lett.*, **7**:118, 1961.

Gordon, J. P., H. J. Zleger, and C. H. Townes: Molecular microwave oscillator and new hyperfine structure in the microwave spectrum of NH_3. *Phys. Rev.*, **95**:282, 1954.

Heavens, O. S.: *Optical masers.* Methuen & Co., London, 1964.

Hellwarth, R. W. and F. J. McClung: Giant optical pulsations from ruby. *Am. Phys. Soc.*, **6**:414, 1961.

Hercher, M.: Design and analysis of Q-switched laser system. *Nerem Record*, Boston, 1964.

Hogg, Christopher and Lawrence Sucsy: *Masers and Lasers.* Maser & Laser Associates. Cambridge, Mass.

Javan, A., W. B. Bennett, Jr., and T. R. Herriott: Population inversion and continuous optical maser oscillation in a gas discharge containing a He-Ne mixture. *Phys. Rev. Lett.*, **6**:106, 1961.

Maiman, T.: Stimulated optical radiation in ruby masers. *Nature*, **187**:493, 1960.

Mann, Daniel P.: Basic aspects of laser operation. *Fed. Proc.*, **24**:8, 1965.

Schawlow, A. L. and C. H. Townes: Infrared and optical masers. *Phys. Rev.*, **112**:1940, 1958.

————: Lasers. *Science*, **149**:13, 1965.

Townes, C. H.: Production of coherent radiation by atoms and molecules. 1964 Nobel Lecture, *IEEE Spectrum,* Aug., 1965.

Weber, J.: Amplification of microwave radiation by substances not in thermal equilibrium. IRE Trans. on Electron Devices *PGED,* 3:1, 1953.

2

Laser Instrumentation

For broad studies of biomedical applications of lasers, including basic research on the laser reaction in living tissue, there is no one single laser which can be used for all these investigative studies. Often as many different lasers as possible should be used for each phase of work. Conflicting opinions are heard about the relative merits of developing new instrumentation or modifying and trying to perfect those lasers now available. Fortunately, there is now a calm reflective period in which those seriously interested in lasers, not in headlines, can plan for specific goals for investigation. For biomedical research, the lasers used are among the solid-state type—ruby, neodymium, and among the gas lasers, argon, carbon dioxide, krypton, nitrogen, ultra-violet and helium-neon.

Some of the many parameters to be considered in the use of laser radiation for biomedical applications, and some of the methods of changing or trying to change some of these variables are as follows:

1. wave length—change of the laser crystal or develop harmonic generation or change to a gas type of laser
2. energy—change by pumping the laser harder and increase the laser input energy or change with the solid laser type to a higher energy laser crystal and more flash lamps
3. density—change by increasing the focus by positive lens or decreasing the focus by negative lens
4. power—kilowatt powers are available from normal mode ruby and neodymium lasers; megawatts powers available from Q-switched mode ruby and neodymium laser and milliwatt and kilowatt powers available from gas lasers.
5. pulse length—method of change is millisecond pulses available for ruby and neodymium laser; approximate range from 0.5–30 milliseconds to nanopulses available from Q-switched lasers; approximate range in most instruments for biomedical applications is 10–80 nanoseconds; continuous wave power would be available from gas lasers.

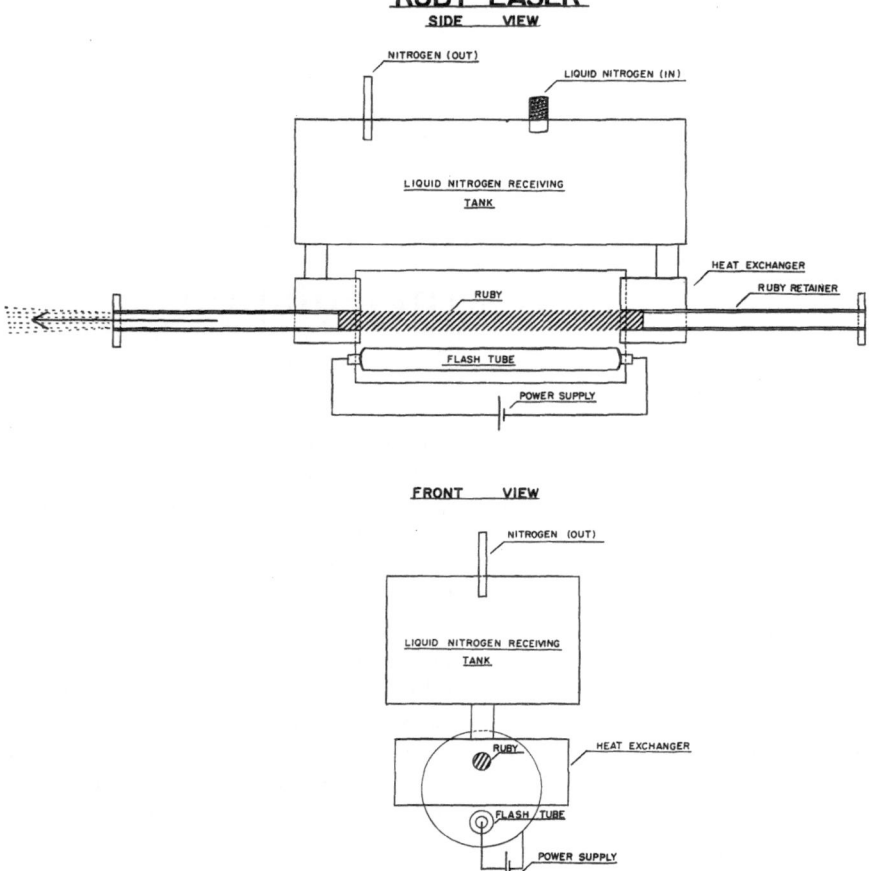

Fig. 2–1. Basic design of old liquid nitrogen-cooled ruby laser.

These variables have to be considered in the use of any one of these specific types of lasers in the biomedical field. All these variables then, even with the same laser, can effect the tissue reactions produced in tissue. Moreover, many laser systems often do not give uniform performance in output pulses. This uniformity and reliance are very important in biomedical research.

For the present, the semi-conductor laser is fixed at too low power levels. Basov claims that with increasing development, a theoretical efficiency of 100% can be achieved with the gas laser. Solid-state lasers have, for the most part, an efficiency of 1–4% and gas lasers an efficiency usually of 0.1% to 8%. Brief mention should be made about the limitations of the electrical energy systems available for high outputs. This

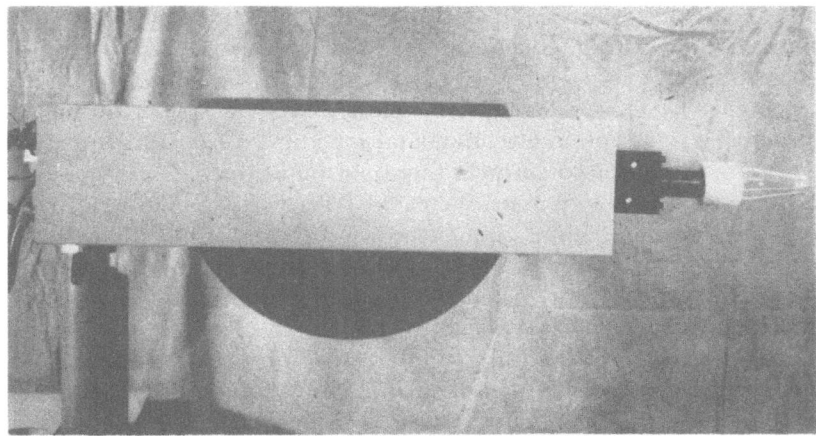

Fig. 2–2. Showing high output, pulsed ruby laser (*Applied Lasers Division, Spacerays Inc.*) with quartz cone to define the general target area and to serve as plume traps.

often means bulky, hazardous capacitor banks. High input voltages from transmission lines can be used also as a source of electrical power. It is hoped that soon chemical processes to produce electrical energy will be available as substitutes for these capacitor banks.

Liquid chemical lasers and plastic lasers are still in early stages of development and not yet available for biomedical applications. One new liquid laser using an inorganic fluid as a light amplifying medium is chiefly a solution of neodymium oxide in highly toxic selenium oxychloride. The wave length of this laser is 10,600 Å. Heller states that if the laser tube breaks, as from shock of the flash system, both the liquid and its vapors are toxic. Continuous wave (CW) operated solid-state lasers and pulse operated gas lasers (except ultra-violet) will not be described since, as yet, these have not been used for biology or medicine.

RUBY LASER

In brief, as indicated previously, the ruby laser is the excitation of chromium elements in a single crystal of aluminum oxide in a reflector cavity. Increasing research in the development of more superior ruby crystals has increased the efficiency of the system. The wave length is in the neighborhood of 6,943 Å and the mode is pulsed. The laser is pumped by flash lamps. In unclassified instrumentation available, exit energies of 1500 joules are under study for biomedical applications. Exit energies vary in ruby lasers attached to microscopes from 0.1 joules to 3 joules. High energy output ruby lasers are also provided with high repetitive rate of discharges giving many impacts a second. In brief, the low outputs are used for investigative work, animal experiments and neurosurgery, while the higher outputs are used for research especially in the field of cancer.

The ruby system is perhaps the most expensive system of the solid state type. At present, a developmental type of rod 5/8 inch in diameter costs more than $300.00 an inch. Increased improvements in the development of ruby laser rods have increased the efficiency of this laser. The laser output will be more symmetrical and reproducible when the mirrors are hard coated on the rods. The detailed techniques of ruby rod selection may be obtained from the literature. High energy ruby systems may cost from $50,000 to $75,000. It is important to note that at times, systems operating at high output do not give uniform performance. Systems are rated usually for level of uniform outputs and level of maximum output. Maximum output may be achieved only a few times.

In an effort to increase the efficiency of the ruby apparatus, liquid nitrogen was used for the cooling system, and dry nitrogen was used to purge the system and to prevent condensation droplets, keep dust away and reduce rate of tarnishing of the silver reflecting cavity. These materials are being replaced by circulating water systems. Rockwell and Hornby indicate that laser rod must be maintained ideally at the same temperature at the commencement of each pulse to get laser outputs which can be repeated. To obtain this, laser systems are cooled with recirculating deionized distilled water flowing around the ruby. The flash lamps and laser cavity are also cooled. The reservoir for the water is maintained at 0°C. In our laboratory, the flow rate is three gallons per minute.

The condition of the reflecting cavity also is very important. Silver reflectance coating is used for ruby lasers, since it has a reflectivity of some 87%, according to Rockwell and Hornby, between 4100 Å and 5600 Å, the absorption band for ruby. At times, flash lamps explode and pit the reflecting surface. This will reduce the output and replating is necessary.

NEODYMIUM LASER

Another example of the solid-state laser is the neodymium laser. Here the rod is glass doped with neodymium. Glass doped with neodymium provides a more economical type of high output solid-state laser. Such neodymium doped glass may be used also for fiber optics. Neodymium fiber optic lasers are now under investigation in eye research and also in dermatology. Neodymium systems do not require any cooling arrangements because of the poor conductivity of glass. Consequently the repetition rate for consistent output firings is not high unless some cooling system is available. Recently, special and more efficient flash tubes have

Fig. 2–3. 200-joules exit energy, air-cooled pulsed neodymium laser. Experimental model of Eastman.

been developed for the neodymium laser. A slope efficiency of 5.1% was obtained by Lesnick and Church, using a coaxial flash lamp as opposed to 3% for a linear lamp. In cancer treatment programs, we have used exit energies of 1160 joules.

The differences in biomedical applications of the ruby, 6943 Å, and neodymium 10,600 Å, will have to be determined by continued research. The wave length output of neodymium indicates that this may have a deeper penetration effect in some living tissues than equivalent energy and power densities of the ruby laser. This has been shown in animal experiments with studies of impacts with this laser over the abdomen. There may be damage to abdominal viscera. Equivalent energy densities for man do not have these deep-penetrating qualities. We have used two high output neodymium lasers from Eastman Kodak Co., one, 200 joules exit energy and the other, 1160 joules of exit energy. The higher output model has been used on a stand mount to increase flexibility of the instrument for biomedical applications.

In general, the neodymium laser is more economical to operate and maintains a more standard output if rapid firings are not necessary. Maintenance costs are also less.

THE Q-SWITCHING TECHNIQUES

The theoretical concepts of the technique of the production of these brief powerful impacts have been described before. The aim is to provide, as Hercher indicates, high peak power by a narrow, angular beam width and a narrow, spectral band width. In biomedical applications, Q-switching has been used both with the ruby and the neodymium. Some

Fig. 2–4. Q-switched ruby laser (*TRG, Inc. #104*) with Daly-Sims attachment for single pulse operation.

of the materials used for the Kerr's cells have included the following: (1) KDP, (2) lithium niobate, (3) copper sulfate, (4) thallium cyanide (5) DTTC-3,3′-diethylthiatricarbocyanide in DMSO-50% efficiency, (6) nitrobenzene, (7) A synthetic garnet, yttrium iron garnet, transparent in the infrared is under study as a material for an infrared laser.

Peak power outputs in Q-switching instruments for biomedical research vary from fractions of a megawatt to 4 and 5 gigawatts. Q-switching with tremendous peak power outputs is very important in the biomedical research, for very little is known about these effects even in tissues as readily accessible as the eye and the skin. Interest in possible ionization effects in the air and in tissue from high peak power Q-switched impacts shows the importance of a continued early study with these types of instruments especially in the treatment of tattoos.

The availability of second, third and fourth harmonic generation from Q-switching continues to open up entirely new fields for biomedical research. Q-switched ruby laser, 6943 Å, produces second harmonics of 3471 Å. With a filter this is passed through KDP crystal and, again with a filter, 1753 Å will be the third harmonics. As yet, there are no x-ray lasers.

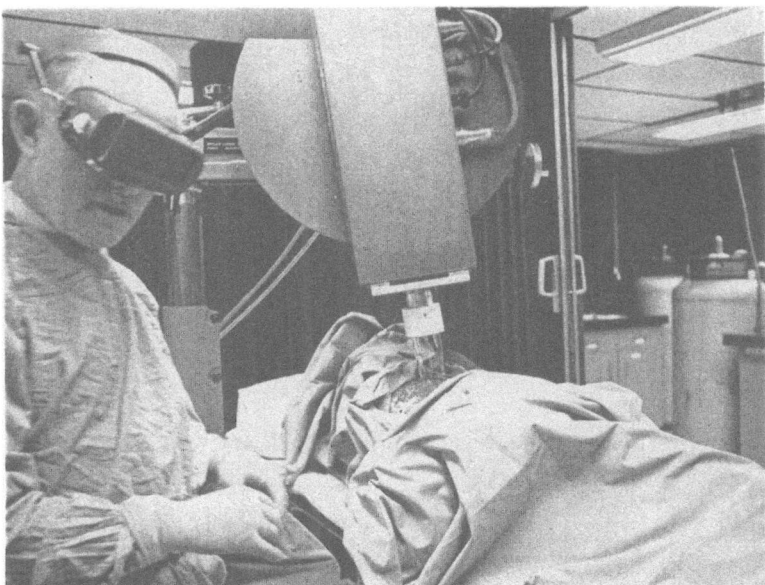

Fig. 2–5. Mobile Laser (*Applied Lasers Division, Spacerays Inc.*) showing position of quartz cone for treatment of melanoma. The quartz cone in contact with the target contains the fragments from the laser plume. These particles include viable melanoma cells. The cone functions as a plume trap.

GAS LASERS

The extensive use of the low output gas laser in the physics laboratory has but recently been paralleled in the field of biomedical applications. This is due to developments of the high output gas lasers, especially the argon and carbon dioxide lasers. Argon lasers have been used in areas where hemorrhage must be controlled and more effectively than its constant rival, the electro-surgical knife. The carbon dioxide laser has no such selectivity.

The argon laser is of special interest and has great promise for laser surgery. Argon ions are excited by electron impact. There are some ten bands of radiation varying from 4545 Å to 5280 Å. Eighty-per cent of the output is 4880 Å to 5145 Å. Mirrors, prisms and mounts attempt to achieve single wave length operations with peaks at 4880 Å and 5145 Å.

The problem of the argon laser is the gas laser tube. The plasma tubes are usually of fused quartz charged with spectroscopically pure gas. The cathode is of an impregnated tantalum to allow for emission at a relatively low temperature. Pinching of the plasma in the capillary tube is done usually in 3 to 5-watt output models by an 800-gauss magnetic field. All the components are water cooled. The tube life varies

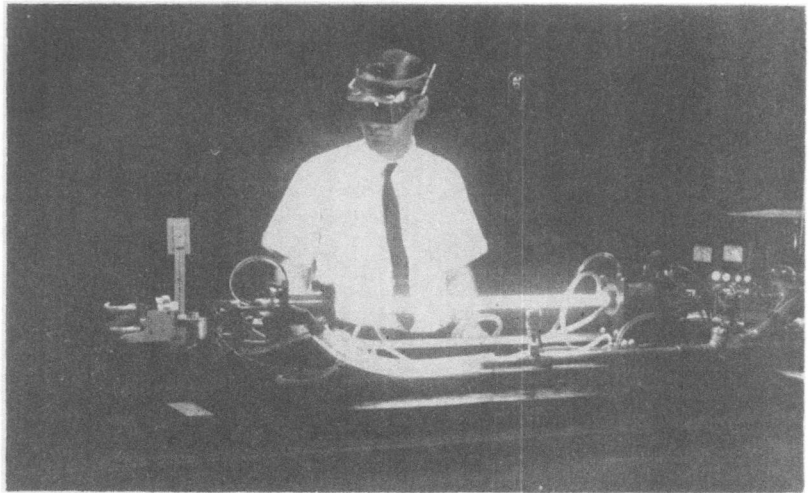

Fig. 2–6. Argon Laser *(Applied Lasers Division, Spacerays Inc.)* .

depending upon its construction and maintained power output. Power outputs now go from less than a watt to 200 watts. For work in laser surgery, power outputs must be in the 1–10-watt range.

The feature of the high output argon laser which is so attractive in the field of biomedical studies, is the continuous wave (CW) output which makes it possible to be used as an optical knife. The absorption of many of these outputs in oxyhemoglobin and reduced hemoglobin of the blood has resulted in special interest in vascular dynamics. Transmission of the argon beam with fiber optics and the hollow, jointed arm probes of Bell Laboratories and spacerays have produced more flexibility than would be available with the reflecting mirror systems.

For the carbon dioxide laser, the output is 106,000 Å. The carbon dioxide laser uses carbon dioxide, helium and nitrogen in its gas tube. The output used for biomedical application can go from 0.7 milliwatt to 24 watts (Perkin Elmer) ; with water cooling to 45 watts. The output can be 50 watts per meter of length. Recently, outputs in excess of one kilowatt have been achieved. The lasing output is said to be about 15%. Because of the frequency of this laser, almost everything can absorb this beam. Special expensive infrared optics are needed with the carbon dioxide laser.

Second harmonic generation from a CO_2 laser has been obtained by Patel with these zinc blend crystals; As, GaS, ZnS, CdTe, ZnSe, ZnTe; and with these hexagonal crystals; ZnS, CdS, CdSe, and a triagonal crystal, Se.

Nitrogen gas lasers are available for biomedical research with outputs

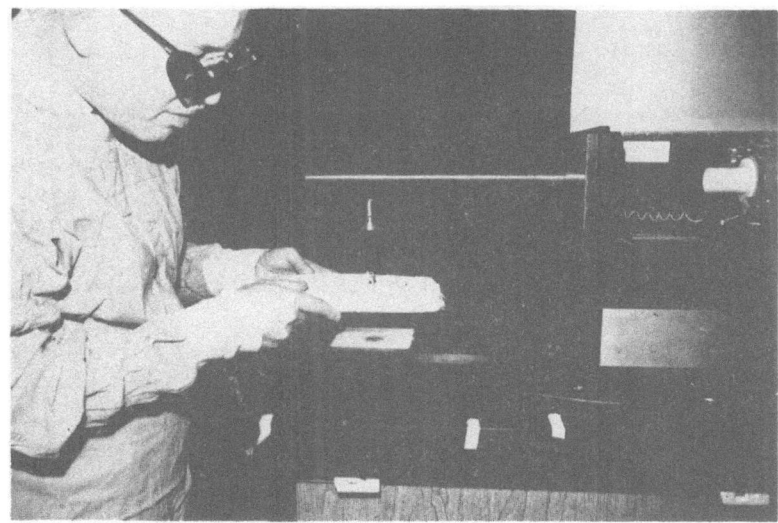

Fig. 2–7. Carbon dioxide laser, 20-watt output (*Perkin Elmer*) reducing a tooth to charred fragments.

around 3371 Å. Such ultraviolet lasers are of importance for work especially with DNA. The output of one of these lasers (Avco) is only 50 KW and they are pulsed. Although this gas laser has a pulse duration of 10–20 nanoseconds and a high repetitive rate of impacts, second harmonic generation from the ruby laser can produce ultraviolet lasing of much higher output. Cheo and Cooper have reported 55 ultraviolet ion laser transitions between 2300 and 4000 Å from the ions of N, O, Ne, Sr, Kr, and Xe. They used pulse excitation. The outputs were low, as have been the ultraviolet outputs from the nitrogen gas laser. Second harmonic generation from ruby gives higher output ultraviolet lasing. Paananen has reported recently, with a current density of 76 amperes, K_2''' continuous wave laser will give 0.3 watts at 3507 Å and Argon $'''$ will give 0.013 watts at 3511 Å. Biologists are waiting impatiently for high output ultraviolet lasing since ultraviolet affects many biological mechanisms. In the field of experimental cancer, ultraviolet lasers will play an important role. Other continuous gas lasers include krypton (5682 Å) and xenon (5971 Å).

The hazard of eye exposure of these high output gas lasers, especially the argon laser, is very important and is mentioned here only to emphasize it. The details will be given later. Because the carbon dioxide laser produces an invisible beam, special protective measures, such as area control and protective asbestos gloves, may be needed.

The helium-neon gas laser, 6328 Å, is used in the physics laboratory for precise measurements. This laser is now receiving increasing atten-

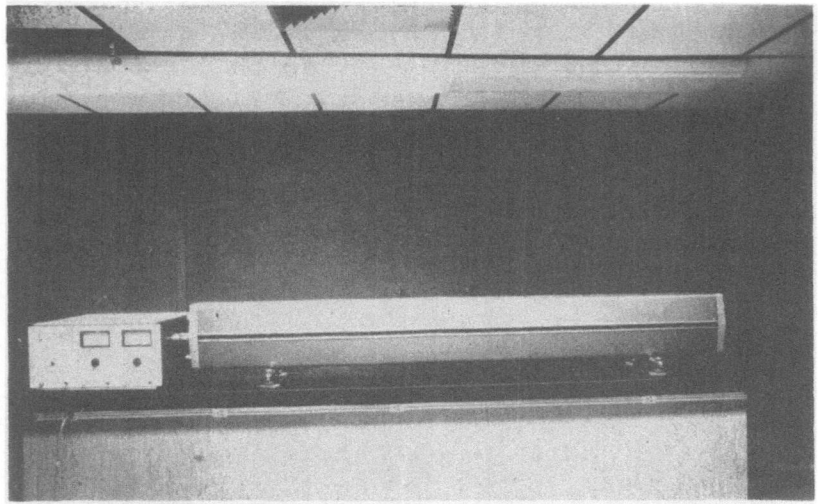

Fig. 2–8. Helium-neon gas laser 75 mW (*Spectra Physics*).

tion in biomedical applications as illumination for photography, including photomicrography, and in the examination of eyes where cataracts may hinder the examination. It is used extensively for holography. Most of the models of helium-neon gas lasers are small portable types, but there is increasing interest in higher power outputs with long helium-neon gas lasers of 250-mW output. Eye protection is absolutely necessary here.

Primarily for use in space communications, a one-watt CW output has been obtained by Young from sun-pumped, neodymium-doped YAG crystals.

RECORDING OF OUTPUTS

Although this feature will be repeated often in the course of this book because of its importance, it is well to emphasize these requirements here. In order to compare data it is necessary to standardize the recording of outputs. Actually, for truly precise work, each output of a laser must be monitored. For this, many parameters have to be controlled. This chapter indicates chiefly the parameter of the laser itself. Some other parameters which also may influence the output of the laser include: (1) its exit energy, (2) its index of reproducibility, (3) the character of its pulse as regards duration and spikes, (4) the lens system and transmission devices which are also used, (5) the target area which is used, including its own particular area and its basic characteristics as regards keratinization, especially its vascularity and connective tissue, and (6) the energy densities which are employed. Finally, there are the parameters of the recording devices to measure these outputs and to record the effects, grossly, microscopically, chemically, cytogenetically, etc.

CONCLUSIONS

From this brief review of the more commonly used lasers in the field of biomedical instrumentation it is obvious that there are indeed many problems. These are expressed often in an apologetic manner by the manufacturer as the "state of the art." Lasers are as yet not mature, standardized assembly-line instruments with specific reproducible outputs. Some of the factors which may be considered as excuses include the poor reliability of the flash tubes, burning of the mirror coatings and imperfection of the crystals. Mention has been made, and will be made repeatedly, of the high costs of laser research, development and instrumentation. In brief, also, until recently interest has not been too great in the biomedical applications because of the relatively limited field of this at the present time, especially with the use of expensive high output instrumentation. Other attractions of laser developments, such as research in weaponry, communications, measurements and calibrations, computers, alignments, ranging, drilling, welding, cutting, wheel balancing and laser holography, are more important for many manufacturers. These are also more remunerative than ventures into the difficult and complicated field of biomedical research. Fortunately, many do cooperate for occasional investigations on animals. We do believe that with properly controlled and planned programs, laser research on man should be expanded to determine parameters of the laser reaction in living tissue and the true applications of laser surgery.

In brief, then, for biomedical research to progress, increasing interest and participation by the laser industry are needed. Instrument development, maturation and decreasing costs are necessary. Most of the attention should be directed to high output CW lasers, especially the argon lasers, but basic research on all lasers must be continued. An important field of biomedical research is the study of laser hazards and the development of laser safety programs with the increasing development of higher and higher output lasers and different wave lengths. Laser technology is now a very important discipline of biomedical engineering.

Table 1

Some Laser Systems for Current Biomedical Applications

Laser	Mode of Operation	Wave Length	Exit Energy Power Output	General Uses	Remarks
Solid-State					
ruby	pulsed (milliseconds) Q-switched (nanoseconds) second harmonics	6,943 Å 3,471 Å	0.1–1500	low energy- laser microscopy animal research eye surgery moderate energy- neurosurgery skin high energy- cancer	cooling system needed
neodymium	pulsed Q-switching	10,600 Å 5,300 Å	10–1160 100 megawatts 4–5 gigawatts	high output- cancer	no cooling system rod can be made very long
Gas					
argon	continuous wave	4,545 Å to 5,145 Å	to 150 watts	surgery in vascular areas cytology research	
helium-neon	continuous wave	6,328 Å	0.001–85 milliwatts	photography transillumination holography	water cooling increases output
carbon dioxide	continuous wave	106,000 Å	0.7 milliwatt-kilowatt	universal tissue destruction	
nitrogen (and other ions)	pulsed	2,300 Å to 4,000 Å	?	biological research	little data available

Table 2

WAVE LENGTH COMPARISONS

Xenon arc	4,000 Å–14,500 Å
Ruby laser	6,943 Å
Second harmonic	3,472 Å
Neodymium laser	10,600 Å
Second harmonic	5,300 Å
Helium-non gas laser	6,328 Å
Argon CW laser	4,545 Å–5,145 Å
	55% = 4,880 Å
	30% = 5,145 Å
	8% = 4,965 Å
Carbon dioxide CW laser	106,000 Å
Second harmonic	53,000 Å
Ultraviolet laser	2,300 Å–4,000 Å

REFERENCES

Basov, N. G.: Semi-conductor lasers. *Science,* **149:**821, 1965.

Cheo, P. K. and H. G. Cooper: Ultraviolet ion laser transitions between 2300–4000 A. Bell Telephone System Technical Publication No. 5030, *Journal of Applied Physics,* **36:**1862, 1965.

Goldman, Leon: Some biomedical engineering aspects of clinical laser surgery. Biomedical Engineering. October, 1966.

Goldman, L., and R. S. Rockwell: The Laser. Advances in Biomedical Engineering and Medical Physics, ed. by Summer Levine, John Wiley and Sons, New York, *in press.*

Heller, A. and A. Lempicki: Liquid Lasers, *Applied Physics Letters,* 9: 106, p. 196.

Hornby, P.: Personal communication.

Lesnick, J. P. and C. H. Church: Efficient high energy laser radiation utilizing a coaxial optical pump. *IEEE J. Quantum Electronics, QE,* **2:**16, 1966.

Paananen, R.: Ultraviolet lasers. *Applied Physics Letters,* **9:**34, 1966.

Patel, C. K. N.: Harmonic generation in IR with CO_2 laser. *Physical Review Letters,* **14:**613, 1966.

Rockwell, R. J.: Personal communication.

———— and P. Hornby: Some aspects of biomedical laser instrumentation. Proc. Symposium on Biomedical Engineering, Marquette University, Milwaukee, Wisconsin, Vols. 1, **199,** p. 323, 1966.

Young, C. G.: A sun-pumped CW one-watt laser. *Applied Optics,* **5:**993, 997, 1966.

3

Laser Output Detection

It is, of course, necessary to know the output of the laser. In the bio-medical field it is much more important to be precise and accurate. The laser reaction in living tissue changes with but small changes in the various parameters of the laser output, whether it be changes in power, energy outputs or pulse duration. The findings of laser research at various centers cannot be compared unless accurate outputs of laser radiation are made. This important field is in a very much confused state at the present time because of the difficulty of measurement of laser output, especially of high outputs. Exit energies above 1500 joules exit energy have been used in biomedical applications with the pulsed solid-state lasers, 0.5–10 watts power outputs are to be used for the argon laser and 25 watts for the carbon dioxide laser. It would seem, at first glance, that the problem of measurements should be simple, for it is merely the detection of classical changes in electromagnetic fields. Many studies of this problem have been done in our laboratory by Hornby and Rockwell.

In general, there are two types of detectors which are available, calorimeters and the photosensitive devices. Recently, because of the increasing importance of the acoustical and pressure wave reactions of the laser, sonic and pressure detectors have been included. Absorbent detectors include copper cone calorimeters, resistance gas and liquid copper sulfate calorimeters of Hornby and Rockwell, bolometers and ballistic piles. The photodetection apparatus includes the photodiode and also the photocalibration associated with magnetic fields. Hornby indicates that the calorimeters of the cone type and the photosensitive type have suffered from the same basic problem since these have to be calibrated from some standard light source. These light sources are of relatively low power, and this in itself introduces the question as to whether calibration of such low power will hold also for the tremendous high power of the laser impacts. Also, there are a number of other factors involved in the calibration of calorimeters so that there is an uncertainty in absolute measurement which is as great as $\pm 10\%$. Moreover, experience in our laboratory concerning the evaluation of current detectors indicates that (1) the measurements may vary from one detector to another, (2) the calibration and setup specifications are not always available, (3) the high energy outputs damage the materials which are an integral part of the calibration system, (4) the calibration techniques using continuous radiators as standards are questionable for the systems in which the pulsed laser outputs are to be measured, (5) the supporting equipment requires periodic calibration as it is an integral part of the final result of calibration, (6) effects of nonlinear scattering are not yet fully understood and may become important factors in high peak power calibration, (7) experimental results in medical laser research are difficult to reproduce from one standard laboratory to another, possibly because the reported outputs are not standardized among the various researchers in the field, (8) periodic recalibration of the laser detectors is mandatory as calibration factors change with use, and (9) the current work in medical laser research is revealing threshold reactions which require the best possible accuracy in measurement.

THE COPPER CONE CALORIMETER

The copper cone absorbing calorimeter is popular in the laser field. In our experience, its repeatability is excellent. The temperature rise in the copper cone may be measured with a fast-acting thermocouple or thermistor. Rockwell and Hornby indicate that the calculation of total pulse energy can be calculated by a knowledge of the absorber mass, specific heat, absorption coefficient and losses from such areas as the entrance window of the copper sulfate liquid calorimeter. The difficult problem here with the device lies in the fact that the aperture receiving the laser beam is too small for work both in medium and high energy laser systems. It becomes necessary, then, to focus directly into the calorimeter. This introduces air factors, namely, reduction of actual beam energy due to reflection absorption of the focussing lens. This leads to inconsistent energy readings because of the variation in focal length of the lens used; according to Rockwell, this variation can be nearly 30%, depending upon the lens. Then, the variation in readings caused by the position of the focal point in the lens with respect to the copper cone aperture can produce variations, at least 10% per centimeter variation, in lens position. Even with the use of a large aperture of the copper cone calorimeter, it was difficult to measure energies above 400 joules at long pulse length. It is also necessary that the detectors, as well as all

Fig. 3–1. Showing set-up for measurement of laser output with TRG copper cone calorimeter and Keithley 150A microvolt-ammeter.

absorbing laser detectors, require 3–5 minutes to elapse between shots for accurate measurements. If it is desired to monitor each impact, and in some circumstances in biomedical research it is necessary to do that, this is too long a period for rapid firing or repetitive firing of lasers.

PHOTOCELL DETECTORS

Photocell detectors are used to measure total energy in conjunction with a resistance-capacitance circuit. Photoelectric detection is used for lasers in the visible and ultraviolet regions. Rest periods between laser impacts are much less than with the calorimeter detection. This is important in the operating room where each impact is monitored and they have to be given as rapidly as possible.

The photodiode with aperture openings of only 0.2 inches is not useful in laser measurements because laser rod diameters of high energy vary all the way from 0.375 to 0.75 inch. If the equipment was not designed originally for laser applications, it is necessary to use diffuser and reduction methods to reduce the light; otherwise, the phototube would be destroyed by the laser. Silicon attenuators are also under study. It is possible, also, that noises, systems from the laser itself, from plasma noise and from mode interference will affect photodiode readings and not give clear-cut pictures. The use of oscilloscopes to detect the energy conversion is a highly sophisticated operation and requires considerable manipulation for the laboratory. The simple fact is that the same type of detectors have, in our experience, a variability of 20% between the different laboratories.

With the use of high energy systems and high peak power outputs nonlinear effects can be observed in the quartz material of the detector and even in some liquids. These interfere seriously with the detection of the output of the laser. As an example, even thumbprints with sebum or sweat on the beam splitters can affect the output because the impact of the laser on these substances can induce second harmonics. This may occur with laser radiation of amino acid crystals.

OTHER DETECTORS

Research continues in the use of the torsion apparatus, liquids, gases, ultrasonic devices, etc. Since the laser impact induces reactions other than thermal in the target, it is necessary to attempt to detect these other reactions. Even radiation detectors, especially of the soft x-ray type, have been used with high power outputs of Q-switching systems.

It is now necessary to monitor high output continuous wave lasers. The argon laser output may be measured with a thermopile or with a photoelectric device. For the carbon dioxide laser, two techniques are used: a water vaporization calorimeter and a silver sulphide absorber connected with a heavy duty calorimeter. The silver sulphide cell may be calibrated from the power output determined by the vaporization calorimeter.

CONCLUSIONS

There is general concern, then, in the field of biomedical application about accurate measurements of laser outputs. Large variations in energy and power outputs reported from the different centers make it impossible at the present time to compare results. In vital structures, such as the brain, heart, blood vessels and eye, accurate evaluation of output is necessary since small variations in output in these vital organs can be significant. It is necessary, then, that more research continue in this field. The description of each laser treatment given to a patient must include the exit energy and energy density of each treatment be recorded accurately. How can this be done, especially in the field of cancer research, when the detectors now currently available are not too reliable?

REFERENCES

Hornby, P.: Personal communication.
Rockwell, J. R.: Personal communication.

4

Light Guides and Techniques of Transmission

It is obvious that for the use of the laser as a biomedical tool, it must come off the optical bench and be flexible. The goal of the laser surgeon is to be able to manipulate the laser beam as accurately and precisely as he does a scalpel. In fact, he would like to call this an optical scalpel.

To attempt to get the laser beam off the optical bench and still maintain its properties of high energy and power outputs, and perhaps coherency, many restrictions are necessary. The usual systems for transmission or reflection for light are available also for the field of laser research. These include, of course, prisms, mirrors, lenses, fiber optics, quartz rods, dielectric tubes and plastics. Many of these have limited flexibility for the laser. The damage to these instruments depends on the energy and power densities used. Recently, gaseous thermal gradient lens systems have been developed to transmit the laser beam. As yet, these have not been applied to biomedical applications.

For all laser systems, it must be remembered that many applications in the biomedical field require transmission of considerable energy and power densities through the various media. These can interfere with the transmission of the beam and can also affect the transmitting system.

PRISMS AND LENS SYSTEMS

High quality prisms can direct the beam off the optical bench. Losses depend on quality, material and size. According to Hornby, losses of 10–15% should be expected. We have used material-type prism systems with exit energy of 1160 joules from the experimental laboratory model of the neodymium laser from Eastman Kodak and have found it effective. Transmission through prisms may not reach inaccessible areas. However, the recent jointed, flexible laser knife probes of spacerays and Bell Telephone Laboratories provide much freedom of movement. Accessory light systems with prisms and with beam splitters are valuable to define the target areas for the laser impact.

High quality lenses have been used to focus the laser beam. These lens systems may be part of the laser beam pathway in the microscope and provide, with corrected lenses, a diameter of only 1 micron for the impact area. With increased power and energy densities from the laser attached to the microscope and especially with the use of Q-swtiched lasers, damage to the lenses has developed. This may involve not only the cement about the lenses, but also the lens itself. A single-lens system, as a substitute for the compound lens, has been suggested to attempt to reduce some of the damage to the glass.

It is desired at times to focus the laser deep into tissues. For example, in studies on impacts of the liver and in some cancers of the neck of patients, we have focussed 3 cm. from the surface. In some animals, we have focussed 4 cm. deep into cancer tissue. In experimental laser neurosurgery, to produce a type of hemorrhage in the brain, a subdural hematoma, it is desired to focus through the skull of animals deep into the brain tissue.

When lens systems are used in laser arrangements, the lenses should be inspected from time to time to observe any clouding. Any defects in the lens may reduce efficiency of transmission and distortion of the pattern. Sapphire lenses have also been suggested for use in the laser system supposedly because of lower absorption for ruby. As indicated previously special infrared optics are required for the carbon dioxide laser.

MIRRORS

Mirrors are a part of the integral system of lasers and also to be used in the transmission of the laser beam. Silver mirrors are used for ruby and for argon lasers. Goldplated mirrors, as developed by the Western Electric Company, are especially valuable in neodymium laser systems as they reflect over 90% of the incident beam with relatively little damage to the surface of the gold mirror, except with high powers. Mirrors also are limited in their use since the failure of alignment through improper manipulation will throw the impact of the laser beam off the target area. In vital areas of the body, this change in target area may be important.

FIBER OPTICS

In recent years, there has been considerable development in the field of fiber optics, both for the physical sciences and for biomedical applications. Significant studies have been done by Kapany, Siegmund and others. Fiber optics transmit light by multiple internal reflections, even through these fibers are bent and curved. Such fibers are available in any diameter of from one micron and up. Intensity losses occur both at the end surfaces and also along the length of the fibers. Fibers may be joined together in bundles. Usually epoxy resins are used to fasten them. Recently, a detailed review of the applications of fiber optics in the field of medicine and in laser biomedical technology of our laboratory and others has been done by J. Goldman, Bereskin, and Shackney. Our experiments with the fiber optics transmission of laser beams have been disappointing since relatively low exit energies will cause considerable damage to the fiber bundle, both to the fibers themselves and to the epoxy resins about such fibers. There is a distinct limitation of the laser transmission in fiber optics. At present, we are investigating the transmission of the argon laser through fiber optics to provide the surgeon with a flexible argon laser knife. Fused fiber optic systems may transmit better than epoxy imbedded fibers.

Also, there is another application of the laser in fiber optics with the use of fiber glass doped with neodymium. In this technique, the fiber optics with neodymium can be lased and a laser beam exits from the end of the fiber optics. This makes available a miniature flexible neodymium laser system. This can be used in the eye and in our current research on appendages in the skin.

We have also used plastics to transmit the laser beam. In our experience even the clear plastic rods do not transmit high ruby outputs without destruction of these light-carrying fibers.

QUARTZ RODS

One of the important current research studies in our laboratory to increase the flexibility of the laser has been with the use of curved quartz rods to transmit high energy laser systems. The exit beam is, of course, divergent. Our original studies were begun with crown quartz glass and have now extended to the use of the more efficient high quality Amersil optical grade quartz rods. These were designed by Rockwell to provide total internal reflection. We have used tapered straight and tapered curved rods with a curve of 42.3% and with exit diameters of 2, 4 and 6 mm. at the tip. The critical angle for quartz to air can be calculated for total internal reflectance in this quality of optical glass with this equation:

$$\sin \Theta = \frac{1}{n_1}$$

$n_1 \quad = 1.4\dot{6}$ (index of refraction of Amersil quartz)

$\Theta c \quad = \sin^{-1} (6.85) = 43.2°$

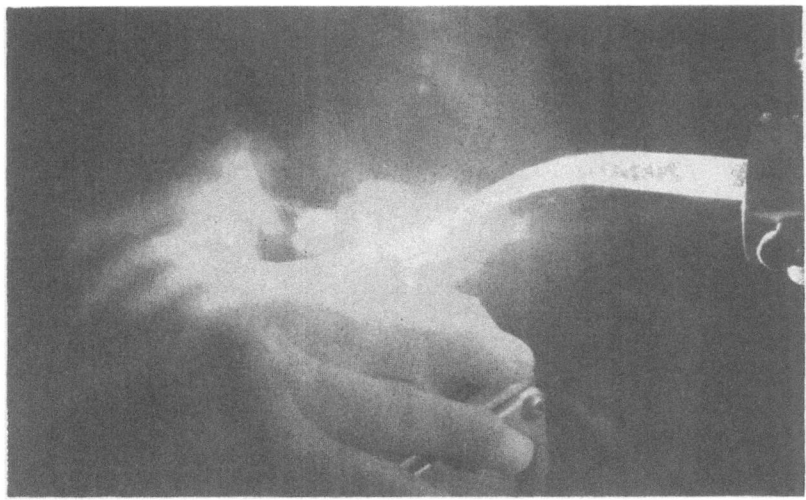

Fig. 4–1. Showing curved tapered quartz rod for treatment of resistant wart about fingernail of physician. Black-and-white copy of colored infrared Ektachrome picture of laser impact *(Eastman)*.

Further studies on end reflection are being done by Rockwell. This means that laser beam rays that are incident upon the quartz to air interphase at angles greater than θ c will undergo total internal reflectance within the quartz rods. Although the changes in the coherency of the laser beam after transmission through these rods have not been determined yet, measurements have been made of the transmitted energy and of the beam divergence at the end. It is of interest to note that the transmission losses in a 6-inch curved quartz rod, 6 mm. exit faces, have been found to be less than 10%. Exit energy from these curved quartz rods have exceeded 450 joules and energy densities up to 2000 joules/cm^2 have been used for target areas such as the inside of the heart of a dog. The walls of the chamber of the heart, the auricle, fastened about the rod near the curvature, showed nothing, indicating there was no escape of the high output beam in that area. The only change in the endocardium was at the exit end of the quartz rod. The surfaces of the rods are kept clean and polished with tin oxide. Polarized light has been used to detect fractures.

The real value of the quartz rods, then, has been to enable the laser to be used in vital areas such as the brain, and to know that the laser is actually on the target even during the respiratory movements of the subject. As indicated, this curved tapered quartz rod has been used about the nose and eye on patients with cancer so that the quartz rods rest directly on the skin, and in operative procedures on patients, into cancerous masses.

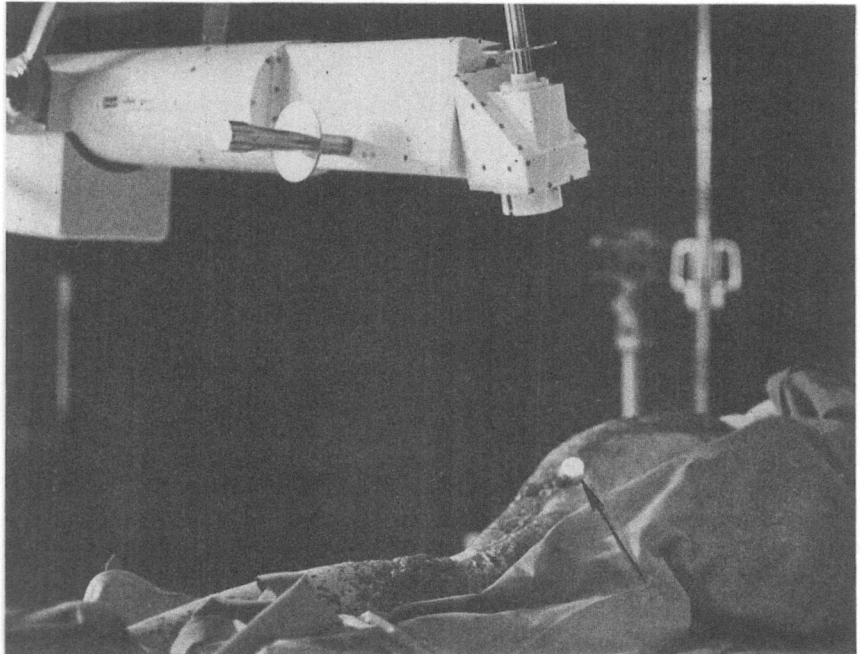

Fig. 4–2. Showing use of bright-spot target light of experimental neodymium laser head (*Eastman*), to define impact area on cancer.

Investigative studies are now being done to combine quartz rods with endoscopic instruments especially with the argon laser. In this manner, the laser may be used as a therapeutic instrument in body cavities. The studies are being done in our laboratory in the field of urology by Mulvaney for the treatment of bladder stones and masses.

LASER HEAD MOUNTS

Another way to make the laser more flexible is to provide an attachment for the laser head. This may be used on a stand as the Eastman Kodak experimental neodymium model. Glenn Hardway, Applied Lasers Division of Spacerays, Inc., has devised an operating room laser attachment. The head of the high output ruby or neodymium laser is attached to a ceiling mount over the operating table. Unlike the portable laser head stand, this ceiling attachment can be manipulated directly over the operating surgeon. Plume traps can be attached to the head.

As yet, it has not been possible to make the high output argon laser flexible enough to bring easily into the operating room even with the fiber optics attachment. Developments continue in this field especially with the flexible segmented light probes of spacerays and of Bell Telephone Laboratories.

CONCLUSIONS

For biomedical applications of the laser, continued research must be done with techniques which transmit the laser beam easily, safely, and effectively away from the laser head onto accessible or inaccessible target areas. This may be accomplished with plume traps and with the optical instruments currently available for the transmission of light and especially with the use of specially prepared curved tapered quartz rods. Continued research is necessary in the field of fiber optics to attempt to adopt this for laser transmission especially with the argon laser for biomedical applications. Studies on neodymium doped fibers are continuing.

REFERENCES

Bell Telephone Laboratories: Laser Knife. Cincinnati *Enquirer*, May 7, 1967.

Goldman, J., P. Hornby, and C. Long: Effect of the laser beam on the skin: Transmission of the laser beam through fiber optics. *J. Inves. Derm.*, 42:231, 1964.

———, and R. Meyer: Transmission of laser beams through various transparent rods for biomedical applications. *Nature*, 1:892, 1965.

———, S. Bereskin, and C. Shackney: Fiber optics in medicine. *N.E. Jour. Med.* 273:1425, 1965 and 273:1477, 1965.

———; P. Hornby, J. Rockwell, R. Meyer, and T. E. Brown: Investigative studies with quartz rods for high energy laser transmission, *Biomedical Engineering Research, in press.*

Hardway, Glenn: Personal communication.

Hornby, P.: Personal communication.

Kapany, N. S. Fiber optics. *Scien. Amer.*, 203:72, 1960.

Koester, C. J. and E. Snitzer: Amplification in a fiber laser. *App. Optics*, 3:1182, 1964.

Mulvaney, W. F.: Personal communication.

Rockwell, R. J.: Personal communication.

Siegmund, W. P.: Fiber optics: Principles, properties and design considerations. American Optical Co., Southbridge, Mass., 1962.

Vogel, K.: Transmission of high power laser light through tapered dielectric tubes and rods. *Science*, 207:281, 1965.

5

Laser Spectroscopy

The laser attached to a microscope is a valuable accessory for emission spectroscopy. The absorption of energy by matter will promote an equilibrium of orbital electrons, and when this energy is large enough, it gives emission of light. In medicine and biology, emission techniques have not gained great prominence by reason of the complexities of interpretation and the sophistication of the apparatus.

There is another technique of spectrophotometry wherein one starts off with a known measure of light and allows this to be selectively absorbed. In absorption spectrophotometry, the amount of light transmitted or absorbed by the material at a definite wavelength is compared to a known reference and this provides a quantitative measure of the sample.

EMISSION TECHNIQUE

In emission spectrophotometry, the material is vaporized and excited. It is important to note that the sample is thereby destroyed. Then the light emitted by the excited materials is analyzed spectroscopically and then photographed. The light emitted by the excited vaporized material is passed through a spectroscope and its characteristic spectrum analyzed. To obtain maximum resolution, the slit for the spectroscope can be varied as well as exposure time, grating adjustment, etc., in order to secure clear photographs. According to Bereskin, a number of difficulties arise in attempting to get good spectrophotographic results. If not enough of the spectrum is obtained in the first analysis, then a number of additional analyses may have to be taken. If this means different samples then there may not be homogeneity of the results. The film has to be of high sensitivity in order to secure good photographs.

In emission flame photometry, the excited material is emitted in the form of light as it is raised to a higher energy state by the flame. The wavelengths of light emitted are characteristic of the particular element and can therefore identify the element. The amount of light emitted is also proportional to the concentration of that particular element.

ATOMIC ABSORPTION

In atomic absorption spectrophotometry, the light beam emitted is directed into a monochromator and onto a detector. As light is emitted, this light is directed through the flame into a monochromator and then into a detector. The detector measures the final intensity of the beam. The amount absorbed by the flame is proportional to the concentration of the sample. Because each metallic element has its own absorption wavelength, a different lamp source must be employed for each element. As a rule, the elements difficult to analyze by emission flame spectrophotometry are more amenable to atomic absorption analysis. The difficulty with this technique of emission flame spectrophotometry then is that since it employs as a source only that which emits the characteristic wavelength of the element, a different source must be used for each element.

Ultraviolet, visible and infrared sections of the electromagnetic spectrum have all been used to provide the light source for absorption spectrophotometry. "The interrelationship of concentration, absorptivity, path length (sample thickness) at a given wavelength is defined by the Bouguer-Beer Law." (Beckman Spectrophotometer Bulletin) The analytical data is found by the curve of sample transmittance or absorbance versus the wavelength in comparison with the spectra of known materials. In available commercial spectrophotometers, like the Beckman, the wavelength settings are reproducible to the 0.05 mμ in the ultraviolet region. Wavelength scale readings are accurate to the 0.1 mμ in the ultraviolet and 0.4 mμ in the visible area.

In atomic absorption, the material may be put into solution. The analysis then is simple and specific. Atomic absorption as a spectrochemical technique has grown and continues to grow fantastically.

Ohman has suggested interference techniques to resolve finer details of an image that could be recorded from the intensity distribution only. This technique he believes could be of value in increasing the resolving power of the spectrograph.

ELECTRON PROBE

The scanning electron probe has been offered recently as a substitute for spectroscopy. In this technique, an electron optical system focuses a beam of electrons in a probe with a diameter less than 0.5μ as its point of contact with the surface of the specimen. Electron excitation results at the emission of x-rays characteristic of the elements present in the specimen. The scanning plates in the objective lens of the optical system make it possible to use the probe over the entire surface of the specimen. The specimen current is used to form a reflection electron picture at the surface and provides the magnification of $3500\times$ comparable to that derived from optical microscopy. The picture is employed according to aid electronics to locate points of zones for analysis. The advantage of this type of instrument is that there is flexibility for analysis over a significant area. The apparatus is not a routine tool. It is cumbersome and expensive. The sample must be put in a vacuum. If this sample is not electrically conductive, it must be made so by coating with graphite. The electron probe then analyzes a much smaller area—in biology, a fraction of a cell.

LASER MICROPROBE

The basic requirements for laser spectroscopy are a laser, a microscope for focusing and viewing, a pair of electrodes and a spectroscope to photograph the spectrum. The laser is a Q-switched neodymium laser in the new Jarrell-Ash model with a peak power output of 7 to 10 megawatts. The target area of 50 micron diameter is selected and it is this area which is vaporized. The weight of the sample is approximately 10^{-7} grams. The vaporized sample, according to Bereskin, is excited to $10,000°$ K and further excitation to $80,000°$ K is done by the auxiliary spark passing between a pair of electrodes just above the sample. In the Jarrell-Ash instrument, the voltage between the electrodes is 2 kilovolts. With the Czerny-Turner spectrograph 0.75 m.f./6.3 the analytical sensitivity is increased 10 times.

The laser attached to the microscope then offers the advantages of both spectroscopy and flexibility for analysis of selected areas in living tissue. The tissue is not destroyed. One limitation of laser spectroscopy is that solids must be used; a liquid cannot be used. It is merely splashed about. However, a filter paper saturated with the liquid may be used. If necessary, a frozen block of the liquid may also be used. Even a microscopic section on a slide may be used.

Rosan summarized recently the use of the laser microprobe.

> The laser microprobe as an analytical tool is cheaper, simpler and more sensitive than the electron microprobe, although not capable as yet of the same morphological or analytical precision. It is cheaper, simpler and less sensitive than an activation analysis unit, but capable of more morphological precision. What the laser does is to vaporize instantaneously a bit of tissue at the focus of the microscope objective through which the laser has been

Fig. 5–1. The laser microprobe. (*Photograph furnished by R. Rosan.*)

projected. The resultant plume or plasma may be analyzed directly by emission spectroscopy or be further excited by an electric spark and then spectroscopically analyzed, or may be conducted into a mass spectrometer and analyzed. The majority of work done thus far has been performed on sparked samples examined by ordinary emission spectroscopy. Theoretically, the size of the probe spot is diffraction limited, i.e. could be smaller than one micron. Actually, such small spots present many difficulties, and the present limit is in the neighborhood of 50 micra for soft tissue. Inside such a spot, sensitivities equivalent to 0.02 meq/1 (an absolute limit approaching 10^{-14} mole) may be expected for such biologically interesting elements as magnesium, calcium, sodium, potassium, zinc and copper.

The laser microprobe is thus an approach to the intracellular distribution of inorganic actions. As has been shown by Scribner, Margoshes and Rasberry at the National Bureau of Standards, with even the relatively crude equipment now available, techniques can be used which give precisions of $\pm 5\%$. When in future years, photomultiplier tubes are substituted for photograph emulsions in the spectroscopic read-out process, precision approaching $\pm 1\%$ should be attainable. Since the laser microprobe is essentially an instrument for rapid analysis of all detectable cations simultaneously, one is thus looking forward to a 5-channel or larger instrument capable of recording biochemically useful data from unstained microtome sections, films of blood or body fluids, chromatographic plates and the like at the rate of a set of analyses every minute. Problems involving the distribution of inorganic cations are fundamental in understanding hemeostasis, aging and enzymatic activity, and many other basic phenomena.

Also possible are methods involving the use of inorganic cations as stains, for macro-molecular cell components, with the subsequent quantification of the stained area on a slide by laser microprobe spectroscopy. Later, with the extension of laser microprobe techniques to mass spectrometry read-out, not only will many anions and even small molecules be added to the list of detectable constituents, but also could isotopes of great biological value, such as deuterium and oxygen. The ability to recognize and quantitate cold isotopes in morphologically defined samples at the cellular level needs no further comment; it could quite easily be of biological importance.

Lastly, a neglected area of laser spectroscopy lies in the clinical use of laser radiation. The plume of excited gases which arises during laser radiation of skin lesions, for instance, contains valuable spectroscopical information on the temperature of interaction, and the radiating species which could be responsible for some biological actions. Also there is a dual relationship involved, which goes back directly to the analytical uses for the laser for biochemical purposes. The laser microprobe can be used directly upon the skin of living volunteers, with no detectable pain. To be able to quantitate intracellular cations from the skin of a living person or animal might be desirable in such physiological problems as heat adaptation or acidosis. Unfortunately, the long-term effects of such minute burns are not known. Therefore, much interchange of information is needed between those using the laser as a tool of radiation biology and those using it as a tool of analytical chemistry. The same sort of interchange is needed in developing laser, especially solid rod lasers, with reproducible control of flux and energy delivered. In both analysis and therapy, this control is critical.

Inasmuch as analytical use of the laser in the microprobe unit depends on a high powered source, rubies and similar materials will continue to serve as generators for some time to come. Though it is now practically possible to develop continuous laser probes based on the 5 to 10 watt CW lasers now becoming available, the enormous problems of securing suitable on-line precision must wait further developments. For these purposes, it would appear that mass spectrometry is the only suitable read-out.

Rosan and Glick have analyzed the minerals in the stomach cell to determine the concentration of calcium and magnesium in the surface cells as compared to the deep cells of the stomach. In children, the pattern is that of more calcium and magnesium on the surface cells as compared to the deep cells of the stomach. In adults, this pattern is reversed. The meaning of this is not known at the present time.

Laser spectroscopy has been used on skin research to determine the inorganic elements in skin tissue. In argyria (silver deposits in the skin), laser spectroscopy can be used to determine the location and concentration of the silver in the skin. The precise localization cannot be as small as with the electron probe. The laser microprobe has also been used recently in a study of the calcium deposits in the skin (calcinosis cutis) following local trauma, and for arsenic and gold. The copper levels in Wilson's Disease, especially after chelation, have been analyzed. The laser microprobe has been used also in studies on teeth and bone. Recently studies have been done on hairs to detect and trace elements. This is of value in medico-legal and toxicological studies. This probe has been used on material obtained from the lesions of a neurotic individual who produced burns on her skin with chemicals. The use of laser spectroscopy in the routine clinical laboratory for rapid analysis for Na, K, and Ca offers great possibilities and in the toxicologic laboratory for As, Pb, etc., offers similar advantages. Laser spectroscopy has been used for melanomas of mice.

The analytical laboratory and the research laboratory will continue research on this valuable accessory tool. It has been used also in ceramic works by Ryan, Clark and Ruh. It has even been used by Brech in the detection of pigments in the forgery of art pieces, and elements in Egyptian glass and enamels, all without gross damage to these objects.

The laser microprobe also offers a technique for detailed and controlled investigative study of the laser plume. This ionized vapor mass may be considered really as being in a "plume trap" where its temperature and particle analysis may be on microanalytical levels. Elements of the laser plume which should be studied are free electrons, energized atoms, molecules, temperature, power density, etc., since these are important in the development of safety programs for the plume. A fast spectrograph can analyze the laser plume without the addition of emission spectroscopy.

RAMAN SPECTROSCOPY

Two important non-linear effects of the laser are the Raman effect and the Brillouin effect. Both these concern the passage of laser light through an optical medium. Raman effect, related to absorption and fluorescence, reveals in the scattered light spectrogram some additional lines not present in the spectrum of the incident light source. In Brillouin scattering, intense ultrasonic energy is produced in which acoustic waves interact with light. Raman spectroscopy gives information on molecular structure. The wavelengths of Raman lines can change with that of the exciting radiation. More frequently, it is in the infrared. With a blue laser, using second or third harmonics, Raman spectroscopy could be done in the ultraviolet. Raman spectra will be used more frequently in the rapidly developing field of laser chemistry for studies of chemical structure. In our laboratory, Schooley is studying organic solvents.

CONCLUSIONS

It is obvious that the laser microprobe, as an accessory for emission spectroscopy, increases the range of use of spectroscopy in biology and medicine and brings spectroscopy to living tissues and even down to cellular levels. Continued development will make for more precise quantitative data.

REFERENCES

Beckman Spectrophotometer Bulletins. Beckman Instruments, Inc. Fullerton, California.

Bereskin, S.: Personal communication.

Brech, F.: Personal communication.

————, and L. Cross: Optical microemission stimulated by a ruby maser. *Appl. Spectrosc.*, **16**:59, 1962.

Glick, D.: The laser microprobe. Its use for elemental analyses in histochemistry. *J. Histochem. and Cytochem.*, **14**:862, 1966.

Johnson, F.: Filling in the blanks in the lasers' spectrum. *Electronic,* **82** April 18, 1966.

Ohman, Y.: A suggested method for increasing the resolving power of a spectrograph. *Nature*, **207**:1284, 1965.

Rosan, R.: Personal communication.

————: On the preparation of samples for laser microprobe analysis. *Appl. Spectrosc.*, **19**:97, 1965.

————, F. Brech, and D. Glick: Current problems in laser microprobe analysis. *Fed. Proc.*, **24**:126, 1965.

———— and D. Glick: Laser beam analyze minerals in stomach cells. *IEEE Spectrum,* July, 1965.

————, M. Healy, and W. McNary: Spectroscopic ultramicroanalysis with a laser. *Science,* **143**:236, 1965.

Runge, E. F., R. W. Mink, and F. R. Bryan: Spectrochemical analysis using a pulsed laser source. *Spectrochem. ACTA,* **20**:733, 1964.

Ryan, J. R., C. B. Clark, and E. Ruh: Laser microprobe analysis of glass defects. Presented at 67th annual meting The American Ceramic Society, Philadelphia, May, 1965.

Schooley, R. E.: Personal communication.

Sherman, D. B., M. P. Ruben, H. M. Goldman, and F. Brech: The application of laser for the spectrochemical analysis of calcified tissues. *Ann. N.Y. Acad. Sci.,* **122**:767, 1965.

Wilson, R., L. Goldman and F. Brech: Calcinosis cutis with laser microprobe analysis. *Arch Derm.,* **95**:490, 1967.

6

Laser Photography and
Laser Holography

Photography is of interest in laser research not only as a technique of recording the effects of the laser but also of studying the dynamics of the actual impact including plume patterns.

To keep records of the effects of the laser, especially in patients, pre- and post-impact pictures of the target area are taken. A standardized technique of photography should be available, preferably in color, to compare results. In our laboratory, the Kodak Instatech close up Camera (Eastman Kodak) with Kodak Ektachrome EX126 film is ordinarily satisfactory for close-up work.

There are two phases of photography used in the study of laser impact. One of these is the direct photography of the impact site, especially during the brief interval of the impact, in terms of milliseconds or even nanoseconds. The other phase is photography with the laser as the source for illumination. Both phases are important in current laser research for studies of the target area and plume, and also for area and personnel protection in the laser laboratory.

Conventional still and high-speed cine cameras, even as high as a million frames per second, have been used. For the study of the bio-medical applications of the laser, pre-impact, impact and post-impact pictures are done, usually in color. For most of the industrial applications of the laser, such as in welding and drilling, black-and-white pictures are usually taken. However, photography in color presents a vivid picture of the impact of the laser, even in industrial processes. Dramatic photographs have been taken to display showers of molten metal or shattering of diamond particles. For color photography with the laser light as the only source, the picture is usually taken with the shutter of the camera open during the laser discharge and the room darkened. It is

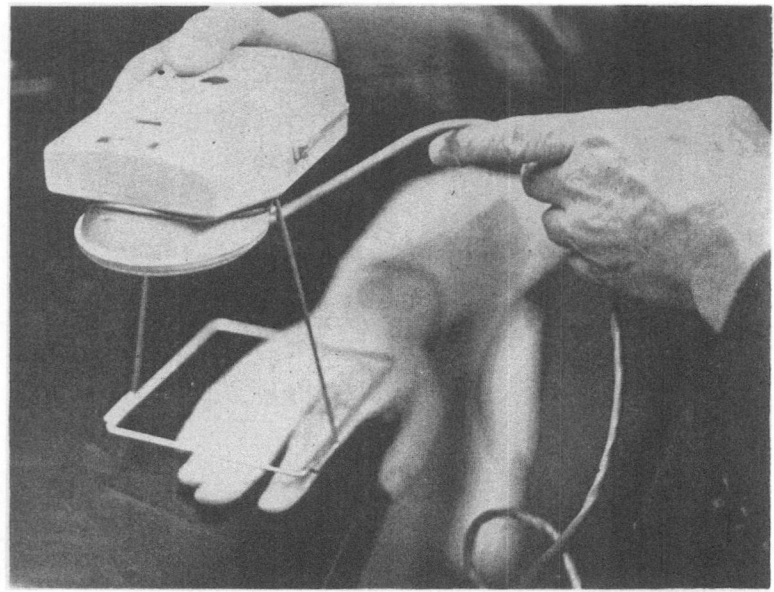

Fig. 6–1. Showing position of camera for photographing tattoo on fingers before and after laser impacts.

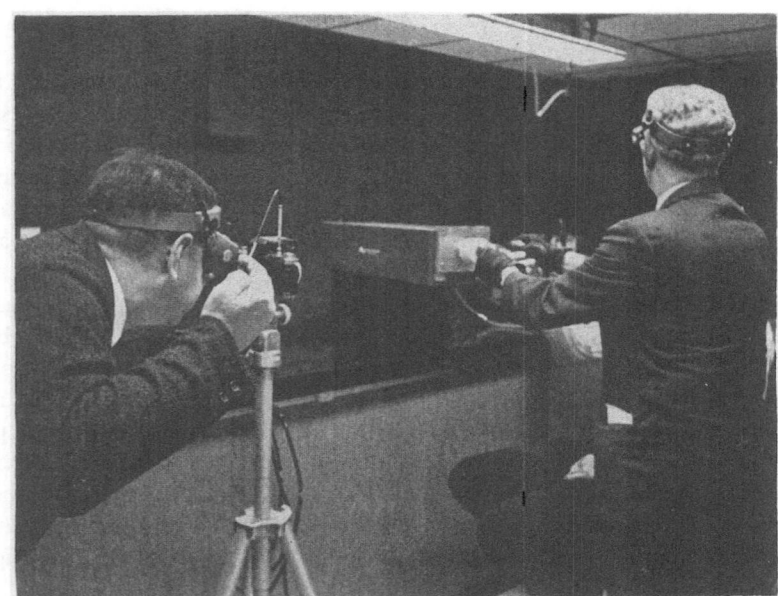

Fig. 6–2. Showing position of wide-angle-lens camera for photographing laser impact.

necessary to add that darkening of the room increases the eye hazards of the laser.

COLOR PHOTOGRAPHY

Color presents, of course, a much more vivid display of the laser impact than does black-and-white photography. Color photography, in addition, is a useful detecting device in studies of second and even third harmonics. Black-and-white pictures do show the laser plume and surface changes in the target area but color reveals these in more detail. Moraites, Owens and Richfield, in our laboratory, list the following exposure factors with Leica and Nikon cameras:

1. If only the impact area is desired, fill the entire frame with the impact area.
2. Telephoto lenses or the 50 mm. close-up lens in the case of the Nikon, are preferred for recording reflection of the laser beam from the target area.

Daylight or Blue Flash Film:

A. 135 mm. lens: distance of 1 meter—f/8—Kodachrome II
B. 90 mm. lens: 20 mm. of extension—f/22—Kodachrome II
C. Nikon with close-up: 50 mm. lens—f/22—10–16 inches

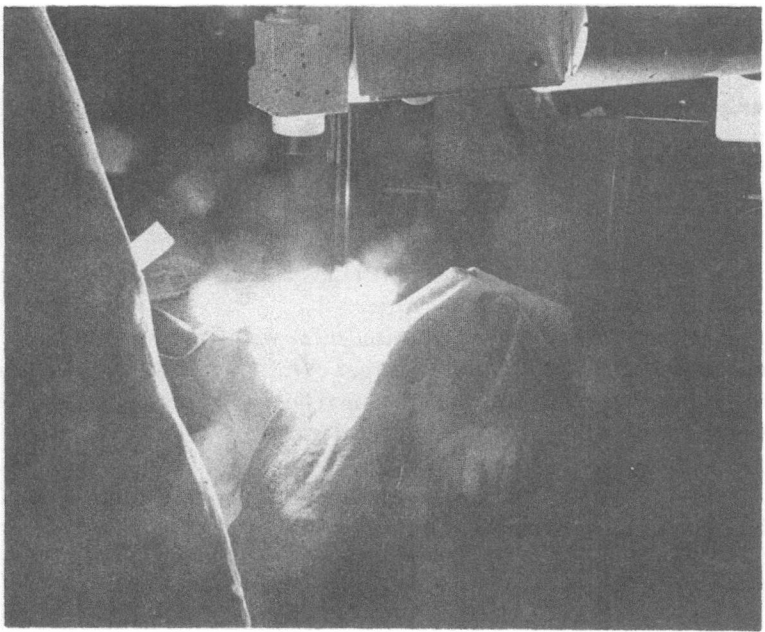

Fig. 6–3. Black-and-white picture of impact of experimental neodymium laser of Eastman 900 joules exit energy unfocussed beam on melanoma of patient.

Color Polaroid pictures of the impact have also been taken.

As yet, we have not observed lens damage with these factors. We have also photographed, in color, Q-switching with peak power outputs up to 100 megawatts, ruby laser, 650 joules exit energies, neodymium impacts on patients at 1160 joules and argon CW lasers to 5 watts.

To show the laser beam in pulsed ruby emissions, condensate from carbon dioxide or, preferably, heavy incense clouds may be used.

INFRARED COLOR PHOTOGRAPHY

For some time we have been studying the use of infrared color film in the laser laboratory under the direction of William R. Buckley, Consulting Dermatologist, Eastman Kodak Company. The film used was Kodak Ektachrome Infrared Aero Film, Type 8443. The particular film which we have been using is a false color material in which green, red and infrared sensitive layers replace the blue, green and red sensitive layers respectively of a normal or natural color film. In this film the infrared images are cyan, green or blue in color. The following exposure factors have been found satisfactory for our work:

ASA to daylight and electronic flash 100.

Exposure must be made with a Wr–12 filter.

Color correction with cc filters is optional—c.c. 10 cyan is recommended with emulsion 8443–15–1.

Exposure latitude is very short and a useable picture will probably be within $\pm\frac{1}{4}$ stop of normal.

Both carbon and silver neutral densities should be avoided unless they are previously calibrated.

Both types of filter shift color balance with density.

Photography is done with the room darkened and with all personnel wearing protective glasses. The photographer focusses his camera. Countdown by the individual firing the laser assists the photographer in determining when to open the shutter.

In our experience, the color infrared film with ruby, neodymium and argon lasers shows details of the target area, the details of the plume and distribution patterns of laser beam reflections about the laboratory. This is true especially with the concomitant use of telephoto lenses. The color intensities give us some idea as to the hazards of the reflected beam from the target area. However, a word of caution is necessary. Because of the distribution of film sensitivities, there is a nearly one-thousand-fold (1,000) differential between the intensities of ruby and neodymium lasers which produce equal film densities. Control studies with black-and white infrared photography have been done, but have been found to be inferior for laser research as compared with infrared color photography.

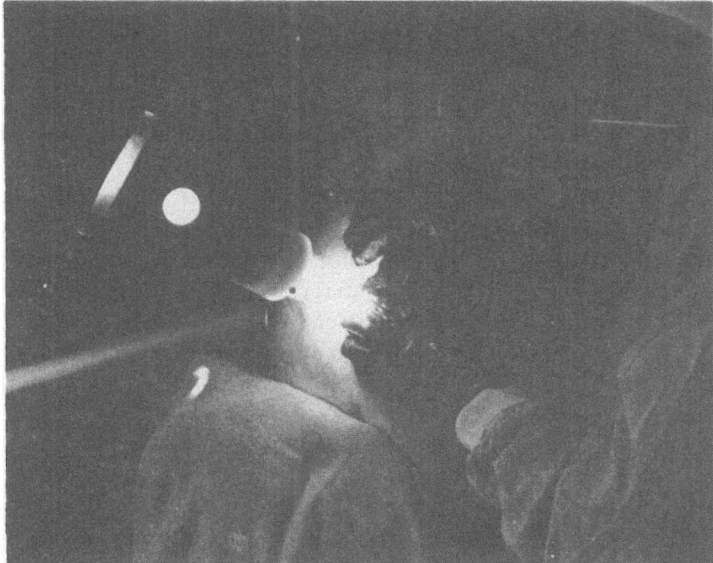

Fig. 6–4. Black-and-white picture of impact of pulsed ruby laser, 1,000 joules/cm² exit energy density, unfocussed beam, on basal cell epithelioma of the back tattooed with red pigment to increase absorption of the laser. Cardboard used to protect adjacent uninvolved skin. Note protective gloves worn.

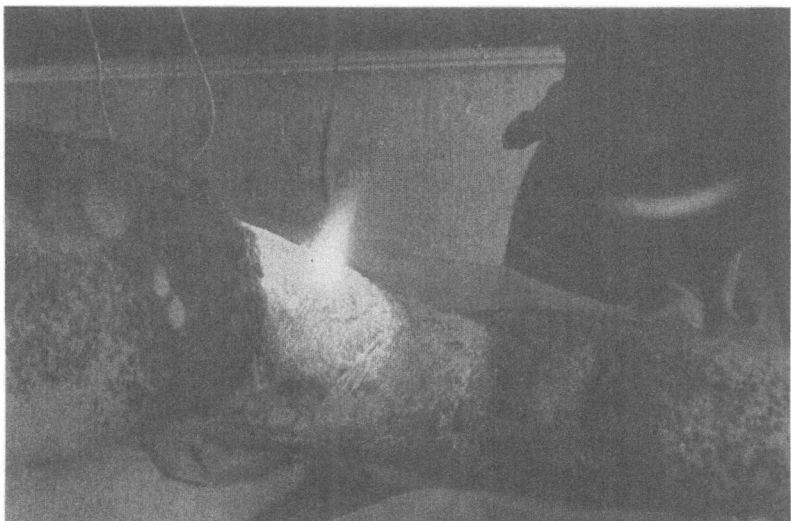

Fig. 6–5. Black-and-white picture of impact of ruby laser, 2,000 joules/cm² energy density, unfocussed beam on metastatic melanoma of thigh; white spots represent areas treated by the laser seven months previously and cleared; with negative biopsies.

Infrared photography is used in our laboratory also for studies of laser transillumination of skin and soft tissues. This technique has been found to be superior to black-and-white infrared photography in showing details of masses or shadow areas. Current studies are in photography, including infrared of laser transillumination of the breast for the early diagnosis of cancer.

Although color photography is of interest in studying the location and the size of the plume of the laser impact, it does not give any help, as yet, in the analysis of the particles and materials and temperatures in the plume. Glass plates and petri dishes placed about the target areas show a type of distribution pattern of the plume fragments. Fluorescence photography has been used to study the efficiency of our plume traps when fluorescein was used in the target area.

HIGH-SPEED PHOTOGRAPHY

In ordinary cine film photography of the laser impact, because of the brief flash of the laser in terms of milliseconds from the pulsed systems and nanoseconds for the Q-switching techniques, little is actually seen. This is true even with the small-diameter beams on lasers attached to microscopes. With high-speed photography (usually around 16,000 frames per second) the impact is more detailed as observed in experiments by Klein and Fine, Minton and McGuff. Local swelling of tissues, described by Kline and Fine as "outward hemispherical distension," and mushroom-cloud effects are seen portrayed especially when impacts are done on animal melanoma, according to Minton. For a more detailed analysis of the laser impact and the plume, the use of higher speeds, even to a million frames a second, should be explored. Analysis of the data may be performed by digital computer techniques.

The scattering of laser rays by small particles has been used by the General Precision Link Group to develop a microparticle laser camera. This camera can measure the size and velocity of extremely small particles travelling at speeds of up to 45,000 miles per hour.

PHOTOGRAPHY OF LASER RODS

Photography may be used also in the inspection of laser instrumentation. One of the many difficulties with laser instrumentation is to attempt to forecast the efficiency of a ruby rod for laser use. In the final analysis, trial and error are the only real methods of testing. However, light transmission techniques may be used to attempt to forecast efficiency. Photography of this light transmission may be taken. The passage of CW helium-neon gas laser light through a ruby rod may sometimes show interference patterns which suggest that the rod is inefficient. Polarized light may also be used for this testing technique. Rockwell has used Mach-Zender interferometric spectrophotometry to study the quality of ruby rod material. Photographs are taken of the interference fringes on the output mirror. Also, in the use of quartz rods for transmission of high energy and high output laser beams, polarized light may be used to determine if stress areas are developing in these rods after use.

LASER HOLOGRAPHY

A recent and fascinating phase of laser photography is the use of the laser itself to develop three-dimensional photography. Holography, invented by Gabor, is a lenseless type of photography using interference techniques. The tremendous popularity of this in the past two years is due to the applications of the laser to this. Coherency of the laser has made this possible. In this technique the laser system uses CW gas lasers to photograph an object. This produces an intricate mass of lines and circles called a laser hologram, a type of diffraction grating pattern. When this negative is placed in the coherent beam of the laser, a third-dimensional image results. If the photographic plate and the objects are illuminated at the same time by the laser beam, the photographic plate can record not only the intensity of the light, but also the phase. Kirkpatrick indicated that the three-dimensional quality "results from its exact reproduction of the light waves scattered from the original objects."

Leith, Kozma, and Upatnieks have listed ideal conditions for hologram construction. These include among other items:

(1) completely stationary components;
(2) monochromacity of the source;
(3) optical flatness of the recording surface;
(4) linearity of the recording process.

Movement of either the reference beam mirror, the object or the recording plate according to these investigators produces image degradation.

In holography, Leith, Kozma, and Upatnieks indicate the photographic film "plays the role of both a square-law detector and a spatial storage device or recorder." Small areas then can store a relatively large amount of data.

Multicolor holograms are possible by using two or more laser beams of different colors to form one single beam. Then this composite single beam is divided into an object beam for the object itself and a reference beam which impacts the photographic plate. The interference patterns of the object beam and the reference beam is recorded in the emulsion of the plate. When this hologram is illuminated by the original laser

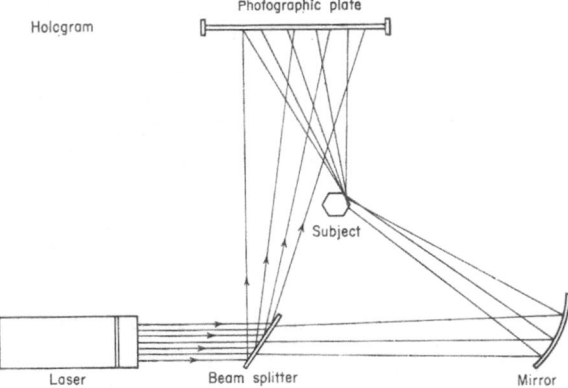

Fig. 6–6A. Showing development of hologram picture. (Robert Meyer-Laser Labora-tory Children's Hospital Research Foundation)

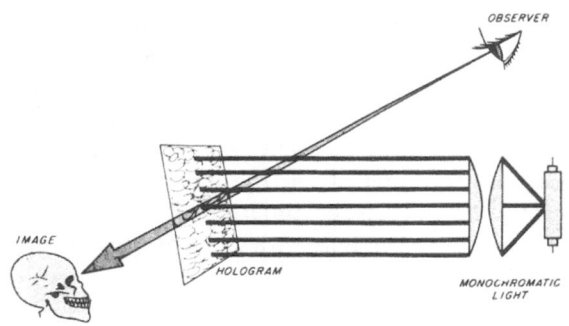

VIEWING A HOLOGRAM

Fig. 6–6B. Viewing of hologram with laser beam. (Ellsworth Cochran-Department of Medical Illustration, College of Medicine University of Cincinnati)

beam, a three-dimensional multicolor image is produced.

Now is it possible to see this three-dimensional color picture by shin-ing white light, from the sun or a flashlight, on this hologram. This has been made possible by the work of Lin and Pennington of Bell Tele-phone Laboratories and Stroke and Labeyrie of the University of Michigan by increasing the angle of the object and reference beams to place interference surfaces in the emulsion closer. This increased num-ber of interference surfaces is supposed to record enough color informa-tion so that even white light can produce this image.

According to Kirkpatrick, holograms have been produced by micro-wave sources and infrared, but as yet not by x-rays. It is obvious how important laser holography would be for diagnostic radiology. Fine and Klein believe that x-ray holography may permit three dimensional pic-

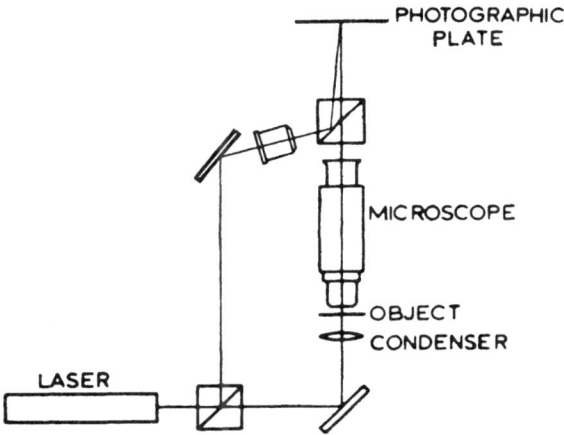

Fig. 6–7. Schematic diagram of holographic microscope. American Optical Company. (Raoul F. Ligten and Harold Osterberg, *Nature* 211:282–283, July 1966.)

tures of macromolecules. Holographic movies may be possible using Eastman Kodak No. 649-F film. The potential uses of this in biomedical applications are evident. A lens-less photography technique has been tures of macromolecules. Holographic movies may be possible using value for fog detection and may have application in studies of aerosols in biomedical research. Holography has been used also to study stresses and strains in metal.

Microscopy by wave front reconstruction using a hologram like a diffraction grating has been done by Leith and Upathnieks. High quality microscopy is obtainable.

An American Optical Company holographic microscope is now available to record third-dimensional data of tissue cultures, protozoans, blood cells, etc. Gabor, Stroke, Brumm, Funkhouser and Labeyrie have recently demonstrated reconstruction of phase objects by holography. This will have great value in microscopy, especially in tissue culture techniques. Wuerker believes that pulsed laser holography may solve the depth of the field problem of the conventional microscope. Upatnieks has used "phase modulating" coherent light to produce better small image holograms useful for microscopy. Holography then by the laser will be used to study effects of the laser at cellular level especially as concerns the picture of shock and pressure waves developing in and spreading from the target area. Holography can be used, then, to measure strain not only in metals, but also in tissues. Holography has been used also in the identification of complex surfaces.

Knox has studied the use of holography in marine ecology. Thurstone has suggested the use of ultra-sonic vibrations to modulate a laser beam to produce a visual picture of internal organs. With this technique of sonic holography he believes that ". . . ultimately excellent images should be realized at reasonable sonic frequencies." This could be done in patients. Sonic and ultra-sonic holography can examine electronic devices for cracks and other defects.

Although helium-neon lasers were used first in laser holography, now the one-watt argon laser can be used (Electro-Optical Systems) with blue-green light. Film exposure time for helium-neon laser holography is up to 15 minutes. For argon laser, the exposure time now is only 10 seconds. As indicated, laser beams of different colors can be combined to form a single beam.

Holography concerns lasers, and so area and personnel safety programs must be included here also. This is especially true during the making of the holograms. If holograms can be viewed in white light and not be the coherent laser, then laser protective glasses are not needed. As yet, the special safety requirement for all phases of laser holography have not been worked out. Van Lighten, Grolman, and Lawton have proposed a holographic fundus camera for research in ophthalmologic holography.

It is evident now that the neurosis called laser hysteria induced by premature unwarranted enthusiasm from arm-chair speculation rather than laboratory work has now affected laser holography.

Table 1
Value of Holographic Imagery (Leith)

1. Third Dimensional.
2. No Blurred Image, Since No Need For Focussing.
3. All Parts Of Object Produce A Sharp Image Regardless Of Depth Of Object.
4. Each Point Of Object Is Stored Over Entire Photographic Plate Instead Of A Single Spot. Damage To A Small Part Of Plate Or Imperfections Will Not Result In Loss Of Any Specific Data Elements.
5. In Some Respects Stored Data Less Affected By A Lack Of Linearity In The Recording Process.

Many more years of detailed work are necessary before multicolor 3-D laser holographic movies, multicolor visual aid medical illustrations of patients are available. Gabor, its inventor, calls many of the current laser holography publicity claims as truly "irresponsible."

RAMAN SPECTROSCOPY PHOTOGRAPHY

The laser as a source for Raman and Brillouin scattering provides for pictures of Raman spectroscopy. Schawlow indicates that the laser for Raman spectroscopy will be useful for studies of material opaque

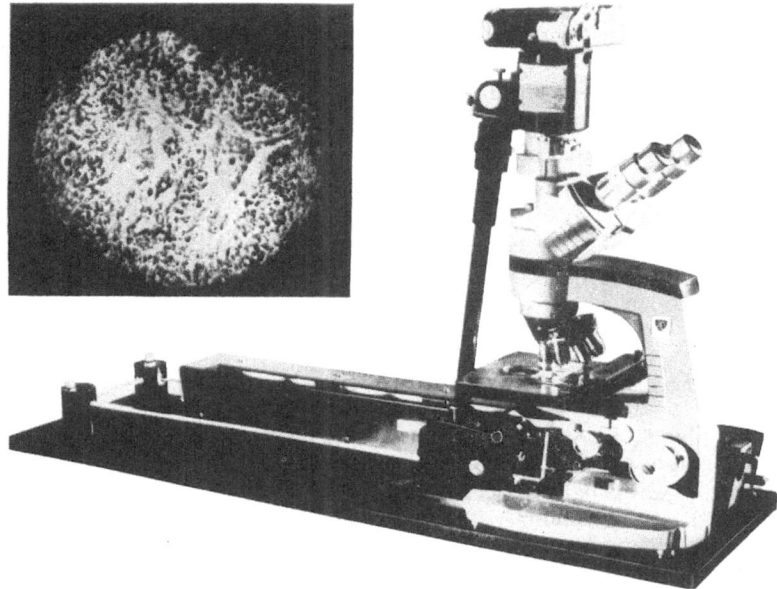

Fig. 6–8. Experimental model of holographic microscope of the American Optical Company, showing three-dimensional microscopic specimen of ????? (*Courtesy, Mason C. Cox, American Optical Company.*)

in the green, blue and ultraviolet ranges. Because of the directional and polarization properties of the laser light, it is convenient for studies of directional Raman scattering and polarization. The gas laser is suitable for Brillouin frequency shifts. These, too, may be photographed.

MISCELLANEOUS

Autonetics Division of North American Aviation Inc. has developed a helium-neon laser system to record photographic images on fine-grain high resolution film with only 1/10,000 watt of light for the picture. The system is said to contain also a laser flying-spot scanner to read out photographs and generate a proportional electrical signal as it scans.

Helium-neon laser light has been used also for photomicrography. The operator should use a filter over the ocular for his eye protection.

For the photographer interested in op art compositions, the laser laboratory provides a fascinating area for investigation and pleasure. These exciting delights should not obscure the continued need for eye protection.

CONCLUSIONS

The spatial and temporal coherency of the laser, its directional property and its great intensity may be used for many phases of photography. The dynamic sequence pictures of the laser impact can be photographed in color and even at high speed. The laser beam itself can be used for laser light photography, high speed photography. It is in the rapidly expanding field of laser holography, especially multicolor, that biomedical applications will be investigated for gross microscopic and eventually animated movie material.

REFERENCES

Bell Telephone Laboratories. Two color pictures produced by lenseless photography. November 12, 1965. Multicolor holograms viewed with ordinary white light.

Brooks, R. E., L. Q. Heflinger, R. F. Wuerker, and R. A. Briones: Holographic photography of high speed phenomena with conventional and Q-switched ruby lasers. *Applied Physics Letters*, 15, August, 1965, p. 92.

Buckley, William R.: Personal communication.

Fishlock, D.: Sound in 3–D. *New Scientist*, 562, Dec. 8, 1966.

Fine, S., E. Klein, W. Nowak, R. E. Scott, Y. Loar, L. Simpson, J. Crissey, J. Donaghue, and V. E. Derr: Interaction of laser with biologic systems. I. Studies on interaction with tissues. *Federations Proceedings*, No. 1, Part III, S–36 (January-February, 1965).

Fine, S., and E. Klein: Biological Effects of Laser Radiation. *Advances in Biological Physics*, New York, 1965, Academic Press. p. 220.

Gabor, D.: Character recognition by holography. *Nature*, 208:422 (October 30, 1965).

———: Holography or the whole picture. *New Scientist* January 13, 1966, pp. 74–78.

———, G. W. Stroke, D. Brumm, A. Funkhouser, and A. Labeyrie: Reconstruction of phase objects by holography. *Nature*, 208:1159 (1965).

Gibson, H. Lou, William R. Buckley, and Keith E. Whitmore: New vistas in infrared photography for biological surveys. *Journal of the Biological Photographic Association*, 33:1 (February, 1965).

Kirkpatrick, P.: Holography. *Laser Focus*, 1:8 (December, 1965).

Knox, C.: Holography for the Marine Ecologist. *Science* 153:989, Abst. *New Scientist*, Sept. 15, 1966, p. 618.

Laser disdrometer.: *Laser Focus*, 1:11 (November, 1965).

Laser Holography.: *Laser Focus*, 7:4–5 (April, 1966).

Leith, E. N. and J. Upatnieks: Microscopy by wave front reconstruction. *Journal of Optical Society of America*, 55:569 (May, 1965).

———, A. Kozma, and J. Upatnieks: Requirements for hologram construction. *The Laser Letter*, 3:2, June, 1966.

McGuff, P. E.: Personal communication.

Minton, P.: Personal communication.

Moncrief, F. E.: The new photography. *Industrial Research* (March, 1964).

Moraites, R. S.: Personal communication.

Owens, P.: Personal communication.

Pennington, K. S. and L. H. Lin: Multicolor wavefront reconstruction. *Applied Physics Letters* (August, 1965).

Schawlow, A. L.: Lasers. *Science,* 149:13 (July 2, 1965).

Stroke, Geo. W.: An Introduction to Coherent Optics and Holography, Academic Press, 1966.

Thurstone, F.: Ultra-sound holograms for the visualization of sonic fields. *Engineering in Medicine and Biology.* Proceedings of the 19th Annual Conference, 1966, p. 222.

Van Lighten, B. Grolman, and K. Lawton: The hologram and its ophthalmic potential. *Amer. J. Optometry and Arch. Amer. Acad. of Optometry,* Monograph 343, June, 1966.

Wuerker, R. F.: Personal communications.

7

Lasers in Modern Chemistry

A recent application of the laser is laser chemistry. In brief, there are three general groups of interest in this field. One of them is the contribution of the laser to the dynamics of photochemistry; next to kinetic spectroscopy; and third to chemical lasers. Photochemistry may include such phases as dye chemistry, especially in solution photochemistry. This dye or colorant research is very important in laser biological research today. Methods of flash photolysis are also of importance. Kinetic spectroscopy would include laser spectroscopy, described previously, and laser excited Raman spectrometry for chemical structure studies. The special studies of laser impacts on DNA, RNA and on enzymes will be considered in Chapter 10. A new world of chemistry has been established by the use of high energy, the so-called "hot atom chemistry." The role of the laser in this field is just beginning. Finally, the energy itself from a chemical reaction has been used to pump lasers.

LASER TYPES

The chemist has at his disposal many lasers of different frequencies with both high and low energy and power outputs. Lasers available are ruby, neodymium and helium-neon gas. Recently high power output CW argon, ultra-violet, and carbon dioxide lasers have become available and can be used for Raman spectroscopy. Then there are the chemical lasers produced by exothermic chemical reactions. One example of this is the hydrogen chloride laser of Kasper and Pimentel. Recently, Kasper and Pimentel have observed laser action during flash photolysis of gaseous CF_3I and CH_3I. Because of the importance of chemical lasers, in September, 1964, a conference was held specifically on chemical lasers. Details of chemical lasers are available from the proceedings of this conference. This report included 21 articles and was 215 pages long.

The making and breaking of chemical bonds releases the energy and consequently Pimentel indicates that a chemical laser could be self-pumping. The chemical action, "once initiated, could proceed without an outside source of power." In the chemical laser, then, the first component of a laser system, the energy level system, is the chemical reaction. The second component of the laser system is the pumping system. As indicated, the energy generated by the chemical reaction produces this. Finally, the third component of the laser system, the optical cavity, is supplied by a quartz tube with laser mirrors. This cavity consists of a quartz tube with quartz end mirrors, tilted "at a carefully calculated angle that minimizes reflection losses in the cavity." Pimentel claims that with this apparatus, about 6% of the radiation in the cavity is deflected to the outside by a flat piece of quartz focussed on a fast-response detector. In addition, for photo-dissociation experiments, he places a flash tube along the laser tube and, as we do with the low energy ruby laser apparatus, concentrates the flash further by wrapping aluminum foil around both tubes. With this setup, with iodine laser emission, Pimentel can obtain power outputs at the kilowatt level.

The chemical laser can give data about the "microscopic distribution of energy in chemical reactions," a type of chemical microscope. Some organic iodides besides CF_3I produce laser emission and some do not. Since many chemical reactions occur with changes of bond lengths and such reactions leave products in vibrationally excited states, Pimentel believes chemically activated vibration-rotation lasers will someday be numerous. These lasers will have to be limited to gaseous molecules with only two to five atoms. They may be difficult to operate in the far infrared. As described previously, a new chemical liquid laser is a solution of neodymium oxide in selenium oxychloride. As yet, these chemical lasers are not continuous wave (CW).

For chemical analysis, as indicated, spectral excitation and photochemical reactions for syntheses and analyses can follow the impact of the laser. As indicated, in Chapter 5, the laser Raman spectrometer has been introduced recently for chemical structure studies.

PROTECTION IN THE CHEMISTRY LABORATORY

The laser chemist must remember, first, that he, like all others working in laser technology, needs protection, even with the low energy lasers. Highly reflectant materials in the chemistry laboratory increase the hazards. With the use of different lasers, different protective glasses must be used. There is no single or universal type of protective glass as yet. The details of such protection of the eye, both from the laser radiation and the laser plume, are discussed in detail in Chapter 21. Laser impact of materials may shatter the test tube or container, especially if these are closed tightly. With the use of high power output

Q-switching techniques, ionization of the air may be produced and such radiation hazards also must be considered. Schooley and Fanney have emphasized the hazards in laser Raman experiments as regards air contaminations with the chemicals and fire hazards.

TECHNIQUES OF EXPOSURE TO LASER RADIATION

For the use of laser radiation for a chemical compound, one concern is the uniformity of exposure of this material. This is especially true for exposure of solutions or suspensions where particle distribution may not be uniform. For example, in the radiation of enzymes and DNA, quartz cuvettes are used, and multiple impacts are given, even overlapping the areas of radiation in order to assure complete, if not uniform, radiation. Light-scattering by the glass container and the possibility of second harmonic generation from the impact on various types of crystals are some other factors to be considered in such exposure experiments.

MISCELLANEOUS REACTIONS

Monomer styrene radiated with a pulsed ruby laser has been converted into a polymer by Yo Han Pao and Peter Rentzepis of Bell Telephone Laboratories. This is among the first specific chemical reactions produced by the laser. With a Q-switched ruby laser, Epstein and Sun have studied the chemical reactions of methane and carbon dioxide. Such high temperature chemical reactions are of value in the study of laser plasmas and in the development of much needed laser dosimetry.

An example of the biologic field of laser photochemical research at the present time is the studies with various frequencies of laser on plant growth, similar to the experiments of Terborgh with the green alga, Acetabularia. The growth in blue light and practically no growth in red light illustrates the differential effects of red and blue on catalase activity, protein and carbohydrate contents and oxygen uptake. Laser impacts on crystalline amino acids may induce second harmonic generation. If this is in the ultraviolet range, color changes, DNA changes, etc., may develop then in living tissue.

The sheet of black silver particles from the silver halide crystals in the gelatin matrix on the photographic film has served as a crude detector of laser impact as regards its energy density, beam width, beam focussing and use as a protection device about the target area, etc. White paper under this impact area from the photographic films shows masses of silver impregnated in it after laser impact. So this film serves as a black body for absorption of the laser energy. Efforts to develop the photographic film as a calibrated detector badge for laser impacts have not been successful. The value of this black film then has been to define the spread of the laser around the proposed target area. The laser has been used for its photographic activity on various photographic film emulsions including the colored infrared Ektachrome (Eastman Kodak). Cholesteric liquid crystals may be used as temperature detectors.

We have studied fluorochromes in relationship to increasing the absorption of laser energy in tissue (so far without success), but we have not studied chemiluminescence. This phase of laser chemistry offers opportunity, as Rounds has shown with his tissue culture studies on the assay of adenosine triphosphate with firefly luciferin. According to Seliger and McElroy, riboflavin, a B-complex vitamin, has been studied in chemiluminescence research.

In this field also, dimethyl sulfoxide (DMSO), a strong hydrogen-bonding organic solvent, has been used. We have used DMSO extensively as a superior vehicle for dye colorants studies in laser research. Because of the toxic properties of DMSO in relationship to lens changes in the eye in some animals and visceral toxicity, the use of this very effective solvent has been curtailed recently in studies with patients.

Laser chemistry may develop new techniques of synthesis especially of macromolecules in polymer research and in biological materials. There is interest in the synthesis of new neutral and acid mucopolysaccharides in the dermis after laser impacts. For space research, heat resistant ceramic materials may also be developed by laser radiation.

As yet, the comparative changes of chemical reactions induced by the laser, and under similar conditions by the plasma torch, have not been studied. This is but another field of research for what may be called loosely, plasma chemistry. Additional studies in space research include the use of laser beams to induce preferential chemical reactions in studies of high energy propellants.

DYES

Dyes are used to increase absorption of the laser beam. Basic chemical structure color type as regards color intensity, solubility, stability and toxicity are but some of the requirements. The changes in chemical structure after impact are also important. This is significant in biomedical applications as regards the possibility of converting a harmless so-called vital dye into a cytotoxic agent, after exposure to the laser. Bleaching of the dye by the laser is also of interest, both for Q-switching techniques and for biomedical applications. Free radical formation after laser impact of methylene blue has been reported by Wiley. The prolonged increase in the signal after laser impact of the dye is significant because it indicates continued free radical production. This also may increase toxicity of the dye by adding ionization radiation within the cell.

CONCLUSIONS

The meager data as yet available in laser chemistry shows that this field is still in its early infancy although it proposes to grow rapidly. The laser chemist now joins the rapidly expanding field of basic laser research with the use of the laser in photosynthesis, chemiluminescence, bioluminescence, spectroscopy and chemical synthesis. The exciting field of actual lasing by chemical reactions has just begun and this too offers much promise for the development of lasers requiring no external source of power.

REFERENCES

Buddenhagen, D. A., A. V. Haeff, G. F. Smith, G. Oster, and K. Oster: Observations of ruby beam intensity patterns with dye sensitized polymers. *Proc. Nat. Acad Sci., USA,* **48**:303, 1962.
Chemical Lasers. K. E. Shuler, Editor.: Supplement No. 2, *Applied Optics,* 1965.
Epstein, L. M. and K. H. Sun: Chemical reactions induced in gases by means of a laser. *Nature,* **211**:1173, 1966.
Fanney, Jr., Julius.: Personal communication.
Goldman, Leon.: Laser in Chemistry. *The Encyclopedia of Chemistry,* 2nd Ed. Reinhold Publishing, New York, 1966, p. 59.
IIT Research Institute Annual Report and Technical Review, August, 1965, p. 11.
Kasper, J. V. V. and G. C. Pimentel: Atomic iodine photo-dissociation laser. *Appl. Phys. Let.,* 5, 1964.
——— ———: Chemical laser. *Chem. and Eng. News,* February 8, 1965, p. 38.
——— ———: HCl chemical laser. *Phys. Rev. Let.,* **14**:352, 1965.
——— ———: Iodine-atom laser emission in alkyl iodide photolysis. *J. Chem. Phys.,* **43**:1827, 1965.
Pao, Y. H. and P. N. Rentzepis: Laser induced production of free radicals in organic compounds. *Appl. Phys. Let.,* **6**:93, 1965.
Pimentel, G. C.: Chemical lasers. *Scien. Amer.,* April, 1966, p. 32–39.
Rentzepis, P. N. and Y. H. Pao: Multiphon proceeds in photochemistry. Abst. 149, *Amer. Chem. Soc. Abst.,* 4:21, 1965.
Rounds, D. E.: Personal communication.
Schooley, R. E.: Personal communication.
Seliger, H. H. and W. D. McElroy: *Light: Physical and Biological Action.* Academic Press, New York, 1965, p. 118.
Terborgh, J.: Effects of red and blue light on growth and morphogenesis of Acetabularia crenulata. *Nature,* **207**:1360, 1965.
Wiley, R. H.: Laser organic chemistry. *Ann. N. Y. Acad. Sci.,* **122**:685, 1965.

8

Basic Mechanisms of Laser Action
on Living Tissue

For those who work in laser technology in fields of laser physics, communication, drilling and welding, it may be of interest to remind them that the effects of the laser on living tissue are quite different. A high energy ruby laser can put a large hole in a copper sheet and then produce only a superficial charring of a large wart on the knee of a patient. Even in the early days of laser research, a focussed beam of a low energy laser was able to make a hole in a razor blade and yet the same energy density had no effect on the normal Caucasian skin of the forearm.

The reason for this difference in response of the target is the fact that living tissue, unlike metal, is complex and heterogenous. It has often a high water content which affects refractive index and heat transference. In addition, there are many different components even in the same layer, each with different physical and absorptive characteristics. All these diverse factors have influence on the absorption, reflectance and transmission of the laser beam. This is important when one plans to focus the laser in depth in tissue, as Hornby as well as Hansen, Fine, Peacock and Klein have shown.

There is a vast literature on the influence of light in biology, but even in this field of photobiology, the laser is a newcomer. So there is relatively little information available on the biologic influences of the tremendous high energy and high power densities of this monochromatic, coherent light beam. Basic studies in this new field of biology are just beginning.

LASER INSTRUMENTATION

In many areas of this book there will be frequent references to the varied characteristics of the laser reaction in living tissues. It is well to

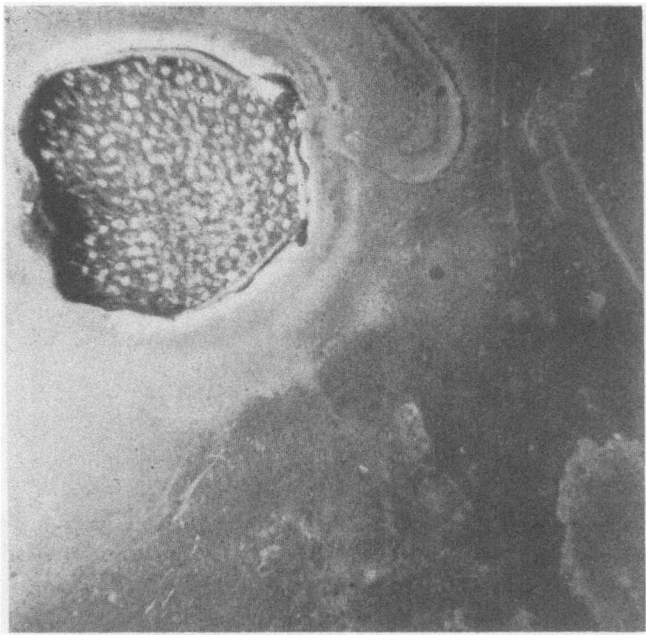

Fig. 8–1. Hole in sheet of copper produced by 198 joules/cm² energy density, unfocussed beam, pulsed ruby laser.

emphasize that these are produced with different types of lasers, including lasers attached to microscopes. This holds also for tremendous outputs now available for the biomedical applications, such as ruby lasers with 1500 joules exit energy; neodymium, at least in unclassified instrumentation, 1160 joules high output; and gas lasers, beyond the kilowatt range.

The basic mechanisms hold also within what has been called loosely "suitable energy range." This indicates that below these energy levels the effects on tissue require absorption of a specific wave length in order to show effects. This is found especially with laser impacts on cells or groups of cells in tissue cultures; as indicated, each target area of living tissue has different optical properties. Yet, for each target there are some basic reactions.

Colored materials in tissue absorb the laser frequency in a single photon process and so produces the biological effects. Rounds and his associates have introduced the two-photon absorption concept to explain the frequency absorption in transparent tissues. Two photon absorption is dependent on high density of the laser and parity levels of the half wave length absorption band.

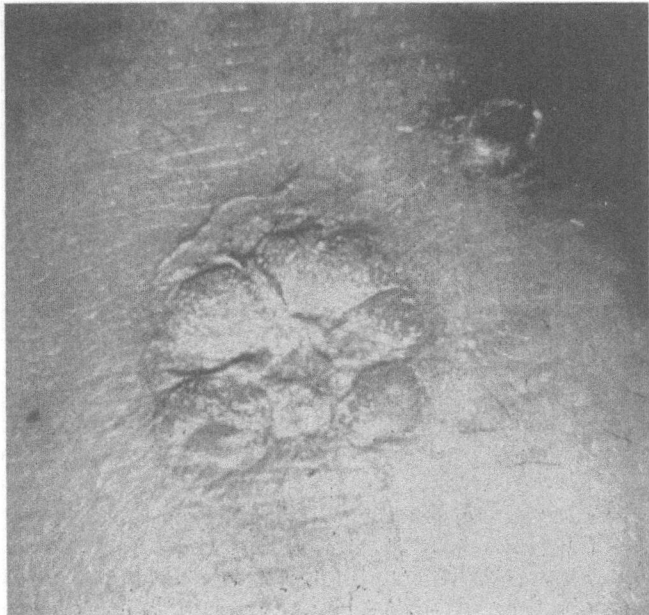

Fig. 8–2. Same energy density on groups of warts on knee, showing only superficial charring reaction. Later, ulceration developed, and wart disappeared with minimal scarring.

REACTION TRIAD OF THE LASER IMPACTS

First, there is a thermal component. Second, there are the pressure recoil and elastic shock waves in tissue. Third, with such high energy densities and high power outputs it is assumed that there must be some electromagnetic field changes.

Some specific phases of this complex reaction are more prominent with certain lasers. For example, the thermal features are especially prominent with normal mode neodymium and ruby lasers. The gas lasers, especially those with high power output, will have strong thermal reactions and the nonlinear effects which intense radiation fields possess, and certainly much less intense with CW lasers. The duration of the effects in tissue of the different phases of this triad reaction may also vary. The thermal phase lasts actually during the duration of the impact although the subsequent elevated temperature of the tissue may persist for some time afterwards. Some parameters of this thermal phase include the type of laser, energy and power outputs, optical characteristics of tissue, etc. It is possible with a focussed ruby laser to have a rise of 75°C from a 100 joule pulse. Possibly the pressure recoil and elastic

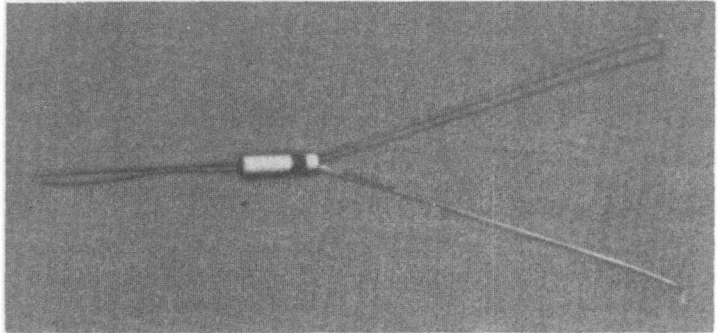

Fig. 8–3. Microminiature thermocouple (*Baldwin Lima Hamilton*).

shock wave effect, and especially the electromagnetic field changes, may be persistent and even progressive. Many more controlled studies are needed.

It is not certain, as yet, whether these different phases have distinct and fixed areas of reactivity in and about the target. Certainly, all three begin in the target site. As indicated, the brief period of impact and the water content of tissue may restrict the thermal factor to a relatively small area unless the thermal factor, as the other, produces thrombosis of blood vessels with subsequent spreading of tissue damage. The sonic and pressure phases may spread a considerable distance from the target area, as Barnes and his associates have shown.

THERMAL

The thermal reaction, from the transformation of laser energy into heat through multiple absorption processes may spread by radiation and/or conduction from the target areas into adjacent areas. As will be indicated in other chapters, measurements of this phase are essentially still crude. In brief, the temperature recorder must have a fast response time, have minimal direct absorption of the laser beam, be miniaturized and strong enough to withstand the impact of the laser. The temperature of the plume of the laser impact may also contribute to the thermal factor. In descriptions of the laser reactions in different tissues, the features of thermal reactions will be described more completely. In controlled studies of the thermal phase, control instruments should include the thermal cautery, electrosurgical units and the plasma torch.

THE ELASTIC STRESS AND RECOIL PRESSURE FACTORS

The need for multiple disciplines to be concerned in laser technology is illustrated by investigations of the elastic stress (ultrasonic) and

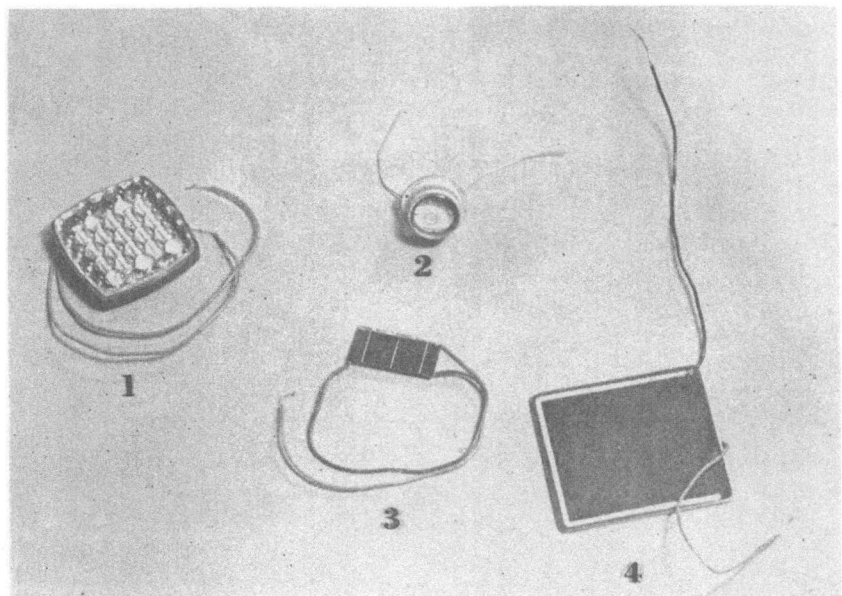

Fig. 8–4. Photodiodes and solar cells: (1) silicon solar cell (*International Rectifier Corp.*). (2) cadmium sulfide cell (*General Electric*). (3) selenium solar cell (*Edmund Scientific Corp.*). (4) silicon solar cell (*International Rectifier Corp.*).

pressure phases of laser impact. This means interest in and knowledge of ultrasonics and the effects of atmospheric pressures of high orders of magnitude. The ultrasonic phenomena developed by the laser are much more powerful than diagnostic ultrasonics. Significant results are produced in tissue by the production of cavitation. Cavitation means the formation of microscopic bubbles through the tissue. In this cavity, gaseous molecules and ionic compounds may be found. This is an interesting phase of the laser impact, not only for the physical changes induced but also from the ultrasonic, and, according to Elpiner, Sokolskaya and Margulis, ultrasonic chemical transformations may develop. This conversion of mechanical energy of sound waves into chemical energy is called sonochemistry by Anbar. This opens a new and promising field for the biologist and chemist. Recently, Wolfgang has reviewed the chemistry of these "hot" atoms produced at high velocities.

Flint has reported shock wave focus pressures over 1×10^6 atmospheres in 50 nanosecond Q-switched pulsing. He has done this with an ultra-fast resolution Schlerien system capable of microsecond interval measurement.

Laser impacts in cavities, such as the eye, brain, chest and bladder, then, are of significance. In these areas the propagation of sonic, ultra-

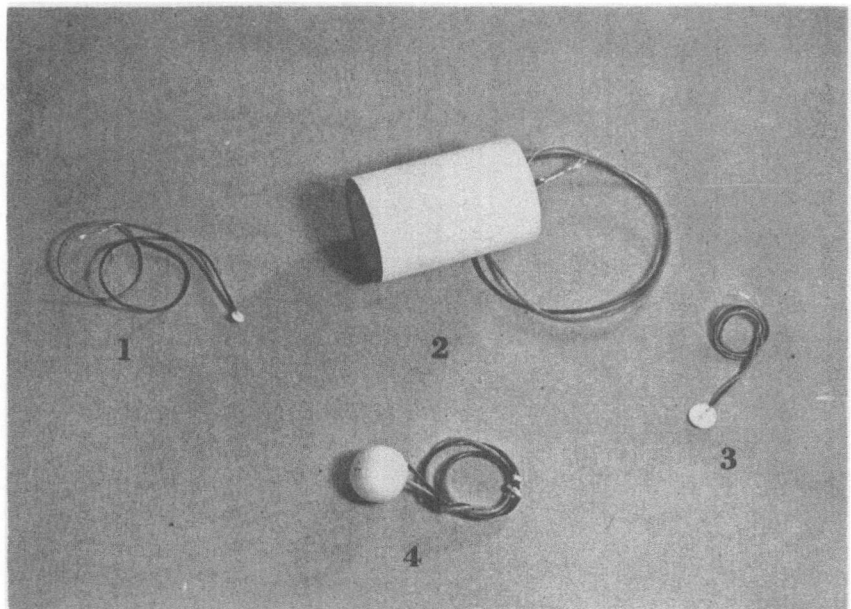

Fig. 8–5. Ultrasonic transducers (*Gulton Industries*) : (1) lead zirconate–lead titanate (2) barium titanate (3) lead zirconate–lead titanate (4) barium titanate.

sonic, and hypersonic waves and pressure phenomena may produce changes distant from the target.

In this phase of the laser chemical reaction, there is considerable interest in oxygen. X-ray radiation effects are enhanced by the use of oxygen. This is true whether the oxygen is used as a local measure, such as hydrogen peroxide, or supplied by more elaborate environmental enclosures, such as the hyperbaric oxygen chambers. Preliminary experiments have been done to provide exogenous oxygen to the tissues also for laser impacts.

ELECTROMAGNETIC FIELD CHANGES

It is said that the electric field strength associated with the high peak power Q-switched laser systems can exceed 10^7 volts per centimeter. This is equivalent to the binding force which holds atoms together.

Proofs of electromagnetic field changes after laser impact have not been many. Heavy particle production is produced by laser impact especially with high peak power Q-switched modes. In addition, free radicals have been detected. The occurrence of free radical formation, especially in pigmented tissues, has been emphasized repeatedly, but Stratton, Pathak and Fine have insisted recently that the laser radiation

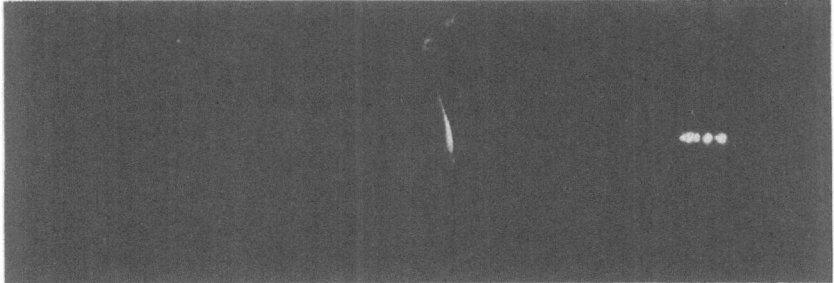

Fig. 8–6. Ionization of air. Q-switched ruby laser, 100 megawatts peak power output. Black-and-white copy of colored infrared Ektachrome (*Eastman*) picture of emission.

only enhances free radical formation from pigmented and melanin colored areas, but not from white skin and white areas. They emphasize that free radical formation differs from that produced by ultraviolet and ionizing energy in tissue. Electron spin resonance (ESR) and electron paramagnetic (EPR) have been used to detect free radicals. So, from these few studies, more research on free radical production from instrumentation presently available is necessary. The fluorescence seen on impact of high power CW argon laser on tissue is not a thermal effect and indicates need for extensive studies.

PIGMENT MASS

All through discussion of the biomedical phases of the laser there will be the emphasis of the significance of the pigmented mass as increasing laser absorption. This will be considered in detail on the chapter on melanin and dyes. Because of its importance it is mentioned briefly here in relation to basic mechanisms. Too little is known about the importance of the distribution of pigment granules, their aggregate or conglomerate collections localized in the nucleus or cytoplasm or even in extracellular zones. It is important to emphasize that the distribution patterns of the natural pigment masses in living cells is probably not the same as that of dye masses introduced into cells in an effort to increase the absorption of the laser by transparent tissues. Smith, in his study of pigment masses in ink, suggests that similar studies be done in relationship to pigment mass distribution in living cells. Some parameters of this problem of pigment masses include their location, size, shape, color and consistency. The triad laser reaction affects these pigment masses. What happens besides vaporization of the mass when it absorbs laser energy? Is there no reflectance from such masses? Are these powerful sources of elastic recoil and pressure waves? Are electromagnetic field changes more prominent here as indicated by ESR and EPR studies? All these are, as yet, unknown.

Transparent tissues may have other processes to effect molecular change from absorption of the laser. As mentioned above, one such process is the two-photon absorption—two photons of the fundamental frequency cooperate in a single absorption event. A second effect in transparent tissue from the strong electrical field is called inverse Bremstrahlung. In this process, loosely bound electrons are accelerated by the strong electric fields, then collisions with adjacent molecules produce thermal effects.

CONCLUSIONS

It is assumed today that a group of reactions develops at the laser impact site and spreads from it. The reactions include a thermal component, ultrasonic or elastic recoil and pressure shock wave phenomena and changes in the electromagnetic fields. All lasers induce such reactions. In some, one or more reactions may be emphasized. Even in accurate rapid measurement of the thermal field, the difficulties of instrumentation are great because of the slow rise of microminiature thermocouples used. Measurement, then, will indicate the relatively slow spread of heating of the tissues, not the initial heat production in the target area. Ultrasonic and hypersonic wave production in tissue and pressure waves are produced locally and spread progressively to produce destructive changes both at cellular and tissue level. Even chemical transformations may be produced; so-called sonochemistry. Tissue ionization in the target area after impact is suspected, but not proved completely. All this means that many more studies of the basic mechanisms of the laser reaction in living tissues should be continued not only at cellular level with the laser microscope, but also in gross living tissues with all the necessary sophisticated instrumentation which these studies require.

Table 1
The Complex Photo-Biological Reactions
With Independent And Collective Actions

Thermal

Pigmented Tissues—high power optical densities
short exposure times (pulsed lasers)
low thermal conductivity

Transparent Tissues—development into an absorbing media by:
"inverse Bremstrahlung"
local heating clue to strong field effects of laser
two Photon Absorption—two photons of the fundamental frequency cooperate in a single absorption event

Table 1—Continued

Elastic Recoil and Pressure Waves
  Electrostrictive forces in a medium of high index of refraction
Electro-Magnetic
  Mechanical and chemical changes through production of cavitation in tissue
  Dissociation and ionization in atomic and molecular fields with free radical
    formation
"Photo-Chemical Reactions"
  A mixture of thermal, ionization and sonochemical reactions
Scattering
  Elastic Photon—particle interactions of Rayleigh scattering
  Photon—Phonon interactions of Brillouin scattering
  Inelastic Photon—particle interactions of Raman scattering

REFERENCES

Anbar, Michael: Chemical reactions induced by sound. *New Scientist,* May 12, 1966, p. 365.

Bach, S. A.: Biological sensitivity to radio frequency and microwave energy. *Fed. Proc.,* 24:22, 1965.

Derr, V. E., E. Klein, and S. Fine: Free radical occurrence in some laser irradiated biologic materials. *Fed. Proc.,* 24:99, 1965.

Elpiner, I. E., A. V. Sokolskaya, and M. A. Margulis: Initiation of chain reaction under an ultrasonic wave effect. *Nature,* 208:945, 1965.

Fine, S., E. Klein, W. Nowak, R. Scott, Y. Laor, L. Simpson, J. Crissey, J. Donoghue, and V. E. Derr: Interaction of laser radiation with biologic systems. I. Studies on interaction with tissues. *Fed. Proc.,* 24:35, 1965.

——— ———, E. Klein, and R. Scott: Laser irradiation of biological systems. *IEEE Spec.,* 1:81, 1964.

Flint, G., quoted by T. E. Gordon, Jr.

Goldman, L.: The Laser Reaction in Tissue, *Laser Cancer Research,* Springer-Verlag New York Inc., 1966.

——— and R. J. Rockwell: Some parameters of Laser Action at Cellular Level. *J.A.M.A.,* 198:173, 1966.

Hansen, W. P., S. Fine, G. R. Peacock, and E. Klein: Focusing of laser light by target surfaces and effects on initial temperature conditions. Presented at NEREM, Boston, November, 1965.

Jacobi, T.: personal communication.

Mendelson, J. A. and N. B. Ackerman: Studies of biologically significant forces following laser irradiation. *Fed. Proc.,* 24:111, 1965.

Rounds, D. E., R. S. Olson, and F. M. Johnson: The effect of the laser on cellular respiration. Presented at NEREM, Boston, November, 1965.

Rounds, D. E., R. S. Olson, and F. M. Johnson: Two-photon absorption in biological materials. *IEEE Quantum Electronics QE* 2441:4, April, 1966.

Seliger, H. H. and W. McElroy: *Light: Physical and Biological Action.* Academic Press, New York, 1965, p. 206.

Smith, F. M.: What is pigment dispersion? Geigy Technical #7, 1965.

Stratton, K., M. A. Pathak, and S. Fine: ESR studies of melanin containing tissues after laser irradiation. Presented at NEREM, Boston, November, 1965.

Wolfgang, R.: Chemistry at high velocities. *Scien. Amer.,* 214: 82, 1966.

9

Physiologic Pigments and Dyes
for Increased Absorption
of Laser Energy

Early in the course of our experiments on the biomedical effects of laser, it was found that the pigmented tissues showed more necrosis from the laser than non-pigmented tissues. This was observed also by Rounds with his studies of laser radiation of tissue cultures and by many observers in laser radiation of animals. Since laser is light, it was expected that dark masses would absorb more laser energy and, as a result, suffer greater degrees of damage. In our studies, colored tissues show more extensive necrosis with ruby, neodymium and CW argon lasers.

THE NATURAL PIGMENTS

The natural pigments of the body for the most part include melanin, having absorption according to Rothman at the violet end of the visible spectrum and no clear-cut absorption bands; melanoid, a pigment having some general absorption at the violet end with a distinct absorption band at 400 millimicrons; oxyhemoglobin with bands at 542 and 576 millimicrons; reduced hemoglobin with a band at 556 millimicrons; carotene with a distinct band at 482 millimicrons; and hemoglobin. All these absorb the laser beam; melanin the most. Even though the melanin granules in normal or cancerous tissue may be found only in the cytoplasm or even in the intercellular areas, the absorption of the laser energy by these granules may destroy the entire cell. At least at cellular level, as observed in tissue culture techniques, cellular destruction is complete. As yet, we do not know clearly what effects the laser beam has as it is transmitted through the so-called transparent tissues. As

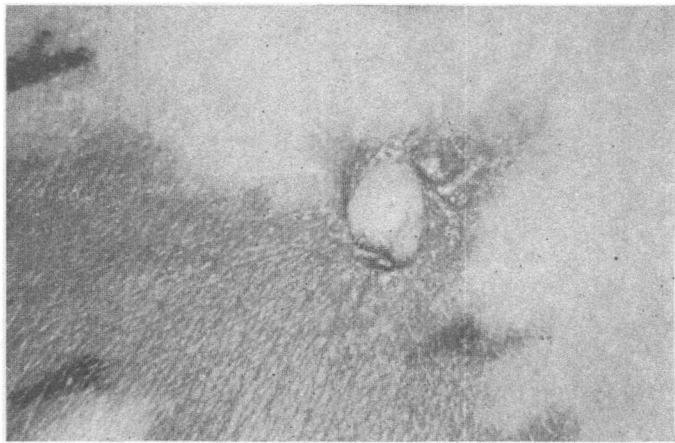

Fig. 9–1. Showing blister reaction of laser impact on colored skin and no reaction in adjacent skin of Negro where color had been lost. Impacted with pulsed ruby laser, 18 joules/cm² energy density, unfocussed beam.

indicated in Chapter 13, apparently the tissue effects are less as determined in studies of white cells, by the changes in motility and functional activity.

In microscopic studies of pigmented tissues exposed to the laser beam, such as pigmented moles and melanomas, a fairly selective action is produced within limits of the energy density available. This is shown by tissue damage more severe in the pigmented areas even though these pigment foci may be distributed in an irregular fashion at the junction between the epidermis and the dermis, and deep in the dermis. This selectivity is more evident in microscopic sections. Often melanoma invades tissue with irregular tumor cords extending in different planes. In melanoma, energy densities even to 1000 joules/cm² still may have some selective effect. With high energy lasers (densities of 5–6000 joules/cm²), there is, of course, some destruction of the transparent tissues between these tumor cords. However, the reaction is more severe in the areas of the pigmented tumor cords.

In types of skin cancer called basal cell epitheliomas, where there is intense pigmentation, the destruction by the laser beam has been much more extensive than in adjacent tumor cords of the same lesion which are not pigmented.

VITAL DYES AND OTHER DYES

These effects observed on pigmented tissues have, of course, suggested the use of dyes for those tissues which are not pigmented. Vital dyes have been used in medical research for many years. A vital dye is, in brief, a type of dye which is absorbed by the cell without any apparent change in the metabolism of the particular cell. In laser dye research Evans blue, methylene blue, and Janus green have been used. Other staining materials such as India ink, Cosmetic black, Sudan Black B, Erichrome black T have not been used as extensively, except by topical application. Dimethyl sulfoxide, dimethyl acetamide and dimethyl formamide and propylene glycol have been used as vehicles for such topical applications. Injections of India ink have been tattooed into soft tissues to enhance the absorption of laser energy. Topical application of this material limits this only to the surface even if such an effective vehicle as dimethyl sulfoxide (DMSO) has been used.

Ritter has listed the basic qualifications of a dye for laser impacts of tumors:

1. To stain broad areas of tumor tissue more than adjacent normal tissue
2. To be non-toxic, at least to normal tissue
3. To enhance laser absorption and thereby to reduce laser energy and power levels to destroy tumor cells.
4. To be able to be given in a suitable vehicle and by a suitable route.

In order to provide for coloring of tissues which do not have natural pigments, the following compounds are among some of the materials which have been used:

1. India ink
2. Evans blue
3. Janus green
4. Methylene blue
5. Acid fuschin dyes
6. Sudan Black B
7. Naphthol blue black
8. Nigrosin W.S.
9. Bromphenol blue used by Goulian and Conway to color necrotic tissue
10. Toluidine blue
11. Erichrome black B
12. Copper sulfate
13. Copper aniline dyes
14. Nile blue dye, sulfate and N-benzyl Nile blue
15. Triphenylmethane acid blue used by Lasarus to indicate border of non-ischemic tissue, a test for tissue viability
16. Cresyl violet
17. Acridine orange
18. Lissamine green
19. Dihydroxyacetone for formation of colored keratin complexes

20. Alcian blue given intravenously by Grant to stain acid mucopolysaccharides
21. Aerosol dye particles covered with bindings

From the listing of these dyes, it is obvious that the range covers both vital dyes and so-called cytotoxic dyes. The purpose of the use of dyes is to attempt to saturate the cell with the dye either through protein binding or by mere diffusion into the cell as foreign masses. Presumably, with the so-called vital dyes, cell respiration and cell metabolism are not interferred with. This can be shown to some extent with tissue culture techniques. For absorption in regions of laser radiation, Klein prefers Janus green, methyl violet, and water black. The acridine group, he believes, does not absorb in these regions. In cancer therapy, intranuclear penetration is desired to attempt to affect DNA. This is important even, for example, in the laser treatment of warts where the DNA dye complex can be affected by the laser and consequently the influence of DNA and RNA on cellular proliferation is affected. With so-called cytotoxic dyes, such as dyes which may be used for cancer chemotherapy, laser absorption and subsequent cellular necrosis may function as an additional cytotoxic effect. In addition, dyes may combine with the amidase enzymes of cancer cells to make for a complex toxic effect to the cell. This would indicate a possible selective complexing by the cancer cell over the normal cell. At present, there is no dye absorbed selectively only by cancer tissue.

As indicated previously in the discussion of laser chemistry, dye chemistry then is an important phase of current multi-discipline laser research. The detailed study of the effect of the laser on the dye moiety as regards the nucleus or its functional group and dye synthesis especially for black dyes are but a few of the parameters.

Dyes are used also in laser technology for Q-switching techniques in the Kerr cell. These cells usually contain nitrobenzene doped with various special solutions that exhibit optical saturation.

TECHNIQUES OF USE OF DYES

In order to provide for penetration of the dye into the cell, various techniques of application have been used. The least desirable is topical application. This often provides only for a surface film, especially with such heavy suspensions as India ink. However, at times this use may be of value. For example, 1% toluidine blue after application of 1% acetic acid selectively stains pre-invasive cancer in the mouth and in the cervix. This can be used for laser therapy then of these pre-invasive areas. This we are attempting to do with the development of the laser colpomicroscope by Schechter. With the treatment of warts by the laser, the dye adjuvants penetrate into the verrucous masses.

Pigments such as red sulphide of mercury have been used by tattoo techniques in miniature piglet and rabbit skin to develop test models for our studies in laser treatment of angiomas.

Laser impact may destroy the dye on the surface and expend all the laser energy without any significant penetration into the depth of tissue. This has been shown by laser impacts on skin of man with India ink, Cosmetic black and tars in various vehicles. Attempts to increase the efficiency of the dye as regards penetration and absorption by the use of highly selective vehicles has not been successful from a practical aspect. As indicated, this includes studies on the excellent vehicle dimethyl sulfoxide (DMSO). Studies with this and related efficient penetrating vehicles are continuing although the use in man of DMSO is limited until more toxicologic data as regards eye damage is available.

The next technique which has been used has been the injection under pressure of the dye directly into tissue. This has been done especially with cancer. The dye has been injected either with syringe and needle or through pressure jet injection techniques such as the Dermojet and Hypospray. Diffuse coloration of tissue usually results. If the laser radiation is to be limited entirely or even partially to a certain area, this type of injection is not effective because of the extensive diffusion of the dye throughout and beyond the target area. With mixtures of dye suspension particles coated with bindings to produce aerosol type of dye, as suggested by Lieberman, it has been much easier to localize the dye depot area. Some aerosol coated dye particles do not dissolve; therefore they do not diffuse. The toxicity of these dye mixtures is not known since only preliminary pilot experiments have been made in our laboratory.

If the dyes are injected intra-arterially, better diffusion into the selected tissue has been accomplished. This has been done by Martens with the injection of methylene blue into the carotid vessel of rabbits for subsequent impact of the vessels of the fundus of the eye. The purpose of this experiment to produce thrombosis was to have selective absorption with lower energy of the laser beam in the blood vessels. This is the experimental background of the use of the laser as prevention techniques for the progressive changes of diabetic retinopathy. Alcian blue has been injected intravenously by Grant to stain acid mucopolysaccharides in venule of the rabbit ear. Laser impacts then produced a thrombosis by selective action in the endothelial or intraluminal side. This dye in his experiments was more effective than Evans blue or carbon black. Dyes have also been injected by perfusion techniques into tumors for their direct cytotoxic effect. These are being used by Ritter and Brown in our laboratory. Nile Blue 2 B has been used in these experiments. This has been done in mammary tumors in C3H/HeJ mice; for selective dye absorption of brain tumors in animals, Nile Blue dyes have been used for their direct cytotoxic brain tumor effect when

injected intravenously. The blood brain barrier makes for good localization into the tumor mass. Dyes may be injected also into lymphatics.
Evans blue can be used here. Then these channels can be impacted by
the laser to attempt to block cancer metastatic spread through the
lymphatics. These techniques offer opportunity for subsequent (and
synergistic?) laser radiation of injected material. This is possible by the
intra-lymphatic injection of radioisotopes complexed with a dye moiety.

MECHANISMS OF DYE RESPONSE TO LASER

The impact of laser on the cell containing pigment, either natural or
induced, often means the vaporization and destruction of this pigment as
indicated previously. Detailed investigative studies are now under way
to study the effect of the laser on the melanin granule. With high
enough energy absorption there is complete vaporization of this mass.
This absorption of photons and subsequent vaporization (laser plume)
may cause cellular destruction through thermal pressure and ultrasonic
factors as suggested in Chapter 8. These changes may spread to adjacent
areas. There is no recognizable morphological or, as yet, histochemical
difference between the laser cell destruction resulting from the use of
natural tissue pigments such as melanin, or those from the use of dihydroxyacetone in keratin or from particles of Evans blue or Janus
green injected directly in tissue. As a rule, with proper energy densities,
the darker the color of the cell, the greater destruction of the cell. Such
studies have been done in melanoma and tattoo in human skin. With
tattoos, even high energy laser systems have been used with exit energies
of 370 joules of neodymium laser, 400 joules of ruby laser, 35 megawatts
peak power output of the Q-switched ruby laser, 2 watts of the argon
laser, and 10 watts of the carbon dioxide laser. In the resultant scar of
such treatment of tattoos, small fragments of the dye particles have been
observed on microscopic sections. In melanoma, as a rule, many melanin
particles are destroyed in toto in the target area. Some remain and are
seen later in macrophages. There is often post treatment melanosis. This
is observed in the retinal epithelium and in melanoma and in its metastases. If there is ideally selective tissue or tumor absorption of the dye,
then, not only will this tissue or tumor be seen more easily, but laser
impacts in the area may destroy selectively the dye absorbent areas.

Clinical experiments have been done with vital dye tattooing and
topical dye mixtures for the laser treatment of multiple basal epitheliomas of the skin, squamous cell epitheliomas, and in the malignant
lymphoma type of tumor. As a rule, the dyes were injected prior to impact. Copper iontophoresis has been done in basal cell epitheliomas.
Topical applications of India ink, Cosmetic black, Nile blue, Nigrosin

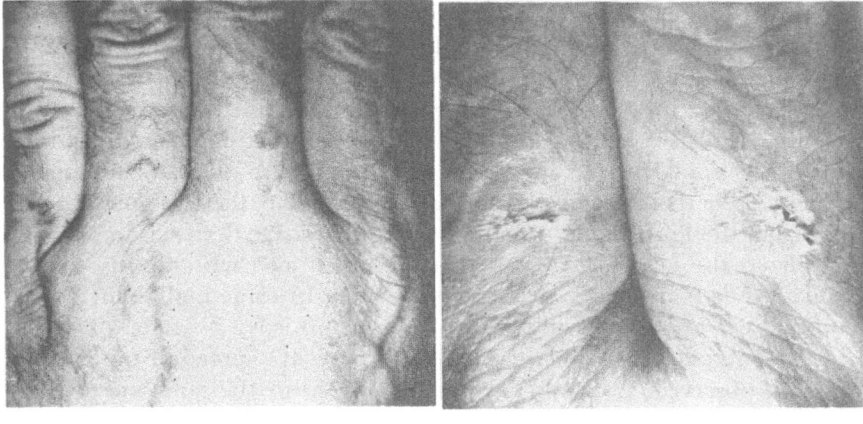

A   B

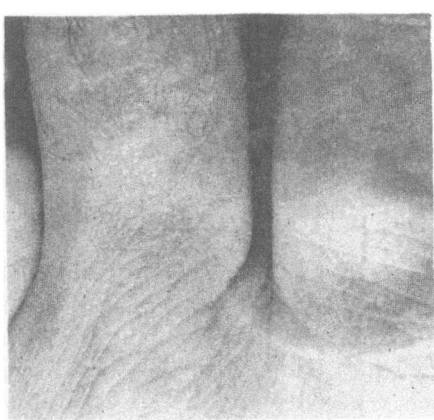

Fig. 9–2. (A) Tattoo of finger before impact. (B) After multiple impacts of 75 joules/cm^2 energy density from pulsed ruby laser, focussed beam. (C) Four months after impact.

C

W.S. or acridine orange and podophyllin have been used in the laser treatment of warts.

In tissue culture techniques, the selective absorption of the dye, and the effect of the laser radiation have been described vividly by Bessis, Malt, and by Rounds and his associates. By the use of high dilutions of Janus green, Amy and Storb have reported selective absorption by mitochondria, and then the selective destruction of these by laser impact.

Progressive research in the laser with the development of different wave lengths will increase the role of tissue colorants. The uses of higher and higher energy systems will cause destruction of tissue even in the absence of color. However, for selective effect adjacent to the target area, dyes will be continued to be used in biomedical applications of laser research. For example, when melanoma is adjacent to vital structures, the laser affects the melanoma, not the underlying major blood vessels or nerves. This we have observed in the treatment of melanoma in the neck. Progressive changes of tissue destruction away from the target area have been shown especially with tissues containing melanin, like melanoma, and nevi which have junctional foci in spotted areas of the epidermis. In junctional nevi with "skip areas" of junctional pigment foci, tissue necrosis spreads selectively to these pigment zones away from the target zone. Whether such progressive destruction of tissue away from the target area in man is observed also with the use of the artificial dyes is not clear at the present time. In some malignancies, the impact areas are usually the width of the unfocussed beam. At times, after laser impacts in pigmented tissues, progressive spread of the necrosis has been observed. Controls have to be done with the same energy densities without the use of the dyes, but with the vehicles alone, and with the use of xenon and other modalities of radiation. Also, to determine the selective effects of dyes where possible, different lasers should be used—ruby, neodymium, continuous wave argon laser—with equivalent power and energy outputs. It is not known at present whether the high output carbon dioxide laser also has color selectivity. At its wave length 106,000 Å, almost every material functions as a black body absorbent.

FIBER OPTICS AND VITAL DYES

In the field of fiber optics microscopy, examination of fresh tissue vital dye staining with Evans blue and Janus green has been shown to be of value in distinguishing different tissue components by Long and also in our laboratory. These dyes are injected into tissue, then the fiber optics probes are inserted, and blood vessels especially are seen in situ after such contrast staining. With fiber optics transmitting laser beams of only low energies at the present, save for the neodymium glass doped fiber, the selective absorption of the dye may be of importance to color tissues so that more selective laser action may be obtained. Tetracycline may be used, for example, for tumor fluorescence and the tissue fluorescence deep in tissue on excitation with Wood's light may be transmitted through fiber optics. It is hoped that with suitable fiber optic systems able to transmit high output argon laser beams, and flexible light probes, vital dyes may be used with argon laser surgery.

CONCLUSION

Color of tissue is important, then, for the absorption of the energy of those lasers which are in use at the present time. To increase the absorption of the laser, non-pigmented tissue may be colored with dye. Some parameters in this field which should be considered are: types of lasers—ruby, neodymium, normal mode or Q-switched continuous wave argon, and carbon dioxide, different wave lengths, the specific dye mixture preferred, the concentration, its methods of applications for the tissue involved. As indicated, advances in laser chemistry, especially for dyes, will aid considerably in the application of this field for laser treatment.

REFERENCES

Amy, R. L. and R. Storb: Selective mitochondrial damage by a ruby laser micro-beam: An electron microscopic study. *Science,* **150**:756, 1965.

Fitzpatrick, T. B.: Mammalian melanin biosynthesis. Prosser White Oration, 1965. Transactions St. John's Dermatological Society, London, 1965.

Goldman, J., P. Hornby, and E. Long: Effect of the laser beam on the skin. Transmission of laser beams through fiber optics. *J. Invest. Derm.,* **42**:231, 1964.

———— S. Bereskin and C. Shackney: Fiber optics in medicine. *New Eng. J. Med.,* **273**:1425, 1965.

Goldman, L.: "Color Qualities of Tissues," *Recent Developments in Cancer Research.* Springer-Verlag New York Inc., 1966.

Goulian, D., Jr. and H. Conway: Dye differentiation of injured tissues in burn injury. *Surg. Gyn. and Ob.,* **121**:3, 1965.

Grant, L.: *The Inflammatory Process.* B. W. Zweibach, L. Grant and R. Mc-Cluskey. Academic Press, 1965.

Klein, E.: Personal communication.

Lasarus, R.: Technique to determine tissue viability. *House Phys. Reporter,* p. 2, September–October, 1965.

Lieberman, A.: Personal communication.

Long, E.: Personal communication.

Martens, E.: Personal communication.

Ritter, E. J.: Personal communication.

Rothman, S.: *Physiology and Biochemistry of the Skin.* Univ. Chicago Press, 1954, p. 519.

Rounds, D. E. and E. C. Chamberlain: Laser radiation of tissue cultures. *Ann. N. Y. Acad Sci.,* **122**:713, 1964.

Schechter, E.: Personal communication.

10

Laser Radiation of Enzyme Systems, DNA and Antigens

Radiation of cellular components, such as enzyme systems DNA and RNA, has been used to study the mechanism of laser reaction in tissue. Such studies not only help to determine how the photons react with living cells, but also whether such radiation is mutagenic or carcinogenic in comparison with other modalities of radiation such as ultraviolet light and x-ray.

SOME ENZYME SYSTEMS

It has been known for many years that enzymes can be inactivated or denatured by various forms of radiation, especially ultraviolet light. In addition, ionizing radiation also causes denaturation or inactivation of enzymes. Radiation can affect the enzyme moiety itself or the solvent with secondary effect on the enzyme. The thermal factor, an important part of the laser reaction in tissue, may affect the enzyme system also secondarily.

The problem of in vitro radiation is not a simple one. Uniform radiation of the solution must be accomplished in order to quantitate the effects. The container must permit the transmission of the laser beam. In our laboratory, we have used Pyrex cuvettes. Klein and his associates have used seven milliliter test tubes with sample volumes of 0.1 to 0.5 ml aliquots. Controls must always be available under the same environmental exposure and time factors. After radiation, the enzyme is tested for activity. Igelman, Rotte, Schechter and Blaney in our Laboratory have studied the effects of ruby and neodymium impacts on trypsin, lysozyme, peroxidase, alcohol dehydrogenase and amylase. Only peroxidase showed any evidence of inactivation. Catalase, which is another

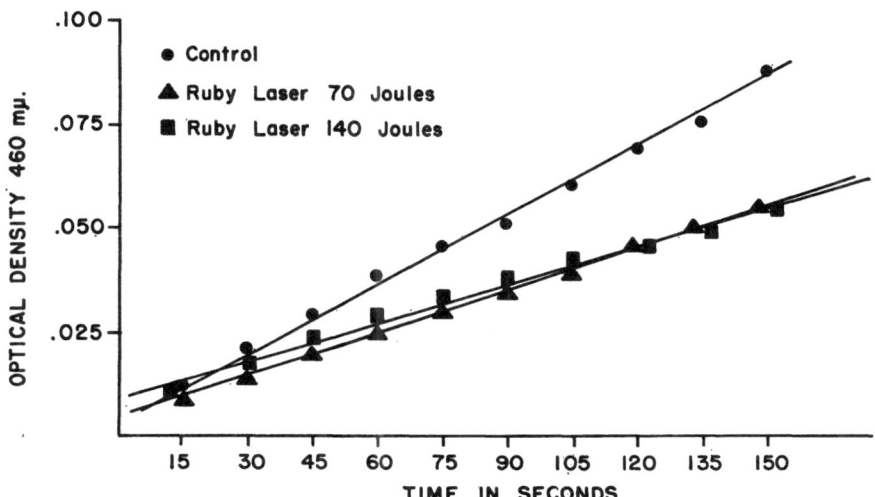

Fig. 10–1. Inhibitory effect of laser radiation on peroxidase. (*Reprinted with permission of Igelman, Rotte, Schechter and Blaney, Department of Dermatology, University of Cincinnati Medical Center.* From: *Annals of the New York Academy of Sciences,* 122:797, 1965.)

hemin-containing enzyme, showed no effects from laser radiation at 6943 Å, even though catalase has been reported to be affected by visible light. Igelman and Rotte also observed no effect of the laser on tyrosinase in solution or in laser-irradiated human skin. Because tyrosine is important in the melanin system, these negative results with tyrosinase are significant. There is no data available yet on the effect of the laser on the tyrosinase inhibitor of Chian and Wilgram. Klein and his associates have studied the effect on solutions of lipase. They found it was not altered unless methylene blue was added. Then the lipase activity was reduced. The lipase used by them also had high levels of proteolytic action. But this proteolytic activity was not reduced by laser radiation whether methylene blue was present or not. They studied also the effect of the laser on plasmin with testing of the fibrinolytic activity, plasminogen activator activity, caseinolytic and TAME esterase activity. Experiments in our laboratory with human plasminogen are detailed in Chapter 13.

Enzymes associated with vascular permeability are important in studies of tissue response to radiation. Jolles and Harrison indicate that enzymatic processes are involved in vascular permeability by experiments using amino-n-caprioc acid as a proteolytic agent. This action "in diminishing the capillary permeability supports the suggestion that this

phase of inflammation is initiated by a protease system liberating proteolytic portions pharmacologically active on capillary endothelium."
These studies can be done with x-ray radiation by observation of the
latent period of local radiation after intravenous injection of vital dyes.
The dyes leak out locally, showing that the blood vessels in the irradiated area are now permeable. With laser radiation, this could be
done with low energy impacts, and in the tissues about the laser necrosis, to detect changes in permeability in adjacent tissues at intervals
after laser impact. In laser eye research, this has been done with fluorescein as a tracer.

DNA AND RNA

Radiation of DNA by ultraviolet light degrades DNA to make it
mutagenic. The effect of radiation of DNA in buffer solution can be
detected by changes in viscosity, temperature profile and midpoint
(Tm) and bacterial transforming principle. In our laboratory, Herr
and Ritter studied preparations of rabbit thymus DNA and also commercial calf thymus DNA. When unstoppered cuvettes were used to
contain the buffer solution for radiation, the evaporation at high temperatures was sufficient to effect the results significantly. Direct thermom-
etry determined the temperature rise in the solution. Experiments with
pulsed ruby laser of 70 joule exit energy produced no significant changes
in DNA. Dyes were not added to the solutions. Electron spin resonance
spectrometry (ESR) studies were also done by Jacobi on DNA radiated
by the pulsed ruby laser in our laboratory and no free radical formation
was detected under the conditions of the first series of experiments.
These are being repeated with high energy systems, Q-switching the
argon laser and the ultra-violet laser.

An attempt was made by Herr also to study bacterial transformation
by using the Marmur procedure for the isolation of DNA from Staphylo-
coccus aureus. A penicillin sensitive strain was used and attempts were
made to transform this into a penicillin resistant strain by isolation of
DNA from the resistant strain, and then after radiation of DNA to
place this irradiated material with penicillin sensitive bacteria. In Herr's
experiments no transformation was accomplished, but the techniques
should be repeated since this is a very significant and sensitive principle
of testing the biological effects of DNA. Klein and his associates tested
effect of laser radiation on DNA by the use of the DNA polymerase
system, reducing primary activity in regenerating rat liver. Rounds and
Adams, with the tissue culture of cells from the carcinoma of the cervix
in women, studied DNA metabolism with amethopterin and adenosine
added to tissue culture to prevent thymidine synthesis. Tritiated thymidine was added and then cultures were radiated with ruby laser; either

G-1 and S-phase of the cell cycle and the DNA metabolism was studied by autoradiographic methods following incubation in tritiated thymidine. A possible site of the action of laser radiation was on the S-phase of the cell cycle. Rounds, Olson and Johnson have also studied the two-photon absorption mechanism on the coenzyme important to intra-cellular biochemical processes, nicotinamide-adenine dinucleotide. Partial inhibition was produced. They conclude that this may be one of the mechanisms whereby intra-cellular biochemical reactions can be affected by the laser.

STUDIES ON ANTIGENS

To determine the effect of laser radiation on other biological systems, tuberculin and histoplasmin skin testing material were radiated by unfocussed beams of 40–50 joules exit energy of pulsed ruby laser. Heat absorption was measured by thermistors in the testing solution. Methylene blue was added to some of the antigen mixtures. These materials were then tested on patients by Prasop of our laboratory. These antigens, with or without dye, showed no changes over the controls in tuberculin and histoplasmin sensitive reactors. Klein and Fine have studied changes in the immunochemical properties of human-globulin at relatively high power levels. Changes in biological activity occurred.

Goldman and Richfield in our laboratory have reported a laser investigator becoming more reactive to the ruby laser after repeated impacts of low energy emission. Attempts were made to sensitize guinea pigs to the pulsed ruby laser. Methylene blue and Janus green were injected into shaven skin of five guinea pigs. In these areas, to attempt to increase antibody production, materials called adjuvants were used. These included the Freund adjuvant, adjuvant 65 (peanut oil, wetting agent, and aluminum monostearate). These were injected, and repeated laser impacts carried out in attempts to produce sensitivity to the laser coagulation necrosis. This was not possible in detailed experiments by Prasop. Controls included laser radiation without the use of these various adjuvants and dyes.

CONCLUSIONS

It is obvious from these preliminary and incomplete studies that the increasing use of laser of high energy and of lasers of different wave lengths, that studies on enzyme systems, DNA and RNA must be continued. Flavinoid enzymes should also be investigated. Because of the importance of lysozymes in inflammation and the effect of the corticosteroids on the laser reaction in tissue, studies on lysozyme should be continued in detail with extraction of lysozyme particles from cells. Controls must include especially the xenon light as well as the ultraviolet light, grenz and x-ray.

REFERENCES

Chian, L. T. and G. F. Wigram: Tyrosinase inhibition, its role in suntanning and albinism, *Science* 155:198, Nov. 14, 1966.

Goldman, L. and D. Richfield: The effect of repeated exposures to laser beams. *Acta Dermato-Venereol.,* 188:773, 1964.

Herr, S.: Personal communication.

Ingelman, J. M., T. C. Rotte, E. Schechter, and D. J. Blaney: Exposure of enzymes to laser radiation. *Ann. N. Y. Acad. Sci.,* 122:790, 1965.

—— and T. C. Rotte: Effects of laser radiation on tyrosinase. *Fed. Proc.,* 24: no. 1, S-94, 1965.

Jacobi, T.: Personal communication.

Jolles, B. and R. G. Harrison: Proteases and the depletion and restoration of skin responsiveness to radiation. *Nature,* 205:920, 1965.

Klein, E., S. Fine, J. Ambrus, E. Cohen, E. Neter, C. Ambrus, T. Bardos, and R. Lyman: Interaction of laser radiation with biologic systems. III. Studies on biologic systems in vitro. *Fed. Proc.,* 24: no. 1, S-104, 1965.

Nitidandhaprabhas, P.: Personal communication.

Rounds, D. E. and J. E. Adams: DNA metabolism in synchronized cells following laser radiation. 3rd Boston Laser Conference, Northeastern University, 1964.

Rounds, D. E., R. S. Olson, and F. M. Johnson: Two photon absorption in reduced nictinamide adenine dinucleotide (NADH) **NEREM RECORD** 1966, p. 158.

11

Laser Applications to Cytology and Cytogenetics

The development of a laser attached to a microscope produced a valuable instrument for work in cytology and cytogenetics. This tool gave the investigator special light beams which could be focussed as accurately as any other form of light and could be placed precisely where it was actually needed. The diameter of this beam has varied from 1 micron up to 100 microns. This has permitted research varying from impacts on the mitochondria to impacts on relatively large areas of tissue for teratological studies on the fertile egg in the chick embryo.

Laser microscopists include Bessis and his co-workers; Amy, Storb and Ter-Pogassian; Chamberlin and Okigaki; Malt; Saks and his associates, Zuzolo and Kopaci; Kochen and Baez; Long; Burbesi; Daniel and Maisel and others.

LASER INSTRUMENTATION

A laser microscope is used. The laser head may be attached to the microscope or it may be substituted for an ocular. It may be attached separately through a vertical illuminator. Finally, it may be completely outside the microscope and with prisms and mirrors reflected through the optics of the scope. This external system, especially employed by Rounds and his associates, has made for increased flexibility, diversity of wavelengths and varying power and energy densities. The limitations are the diameter of the beam and the low energy density.

Both ruby and neodymium lasers have been used; Q-switching has also been employed, especially in microprobes for laser spectroscopy. For most work in cytology, the phase contrast microscope has been used. Time lapse photomicrography may be employed to record the effects of the impact. To compare the effects of coherent and non-coherent radiation, Tomberg, a pioneer investigator in laser biomedical research, has

81

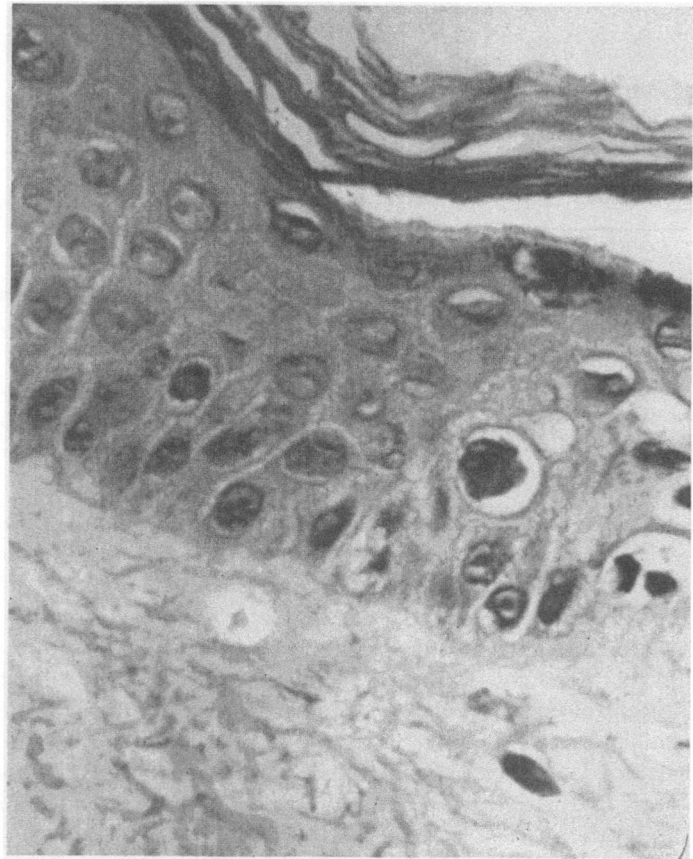

Fig. 11–1. Showing abnormal mitosis after laser impact on normal skin. Seven days after impact of 20 megawatts peak power output from Q-switched ruby laser, skin appeared grossly normal. Hematoxylin-eosin, ×360.

used the inverted microscope. As described previously, now the holographic microscope is also available.

Protection of the operator must be considered in laser microscopy. Some devices employ a metal shield to block off the oculars when the laser is charged and the laser cannot be fired unless the metal shield is in place. Others may use a closed-circuit television. Unfortunately, other set-ups require the operator to be alert and to turn his head away from the impact. It is still necessary to emphasize eye protection for personnel because even with low energies, the reflectant beam from the microscope stage may be sufficient to damage the eyes of the operators.

There should be some concern also about the hazards to the microscope equipment. This is true especially with high power and high

Fig. 11–2. Showing set-up of laser with phase-contrast, time-lapse photomicrography.

energy densities where expensive objectives may be ruined, either the lens itself directly or indirectly from vaporization of the cement around the optics. The optics should be examined from time to time to assess the damage. This is especially necessary with techniques employing Q-switching.

In photomicrography, the choice of film is of some importance. High speed color film is preferred for color work. The colored infrared film from Eastman Kodak described in Chapter 6 has been used to study the patterns of the actual impact and the laser plume. This film is also under study now in photomicrography of fixed sections.

IMPACTS ON RED CELLS AND PROTOZOA

The simplest experiments which can be done with laser microsurgery and cytology are impacts of red cells in a fresh smear on a microscope slide and impacts on paramecium. These studies will show the effect of color on absorption of the laser beam. Additions of minute amounts of dye such as Janus green, Evan's blue or methylene blue may render transparent cells more reactive to the laser. In Chapter 13, studies on the red and white cells are detailed. The studies of Bessis and his associates illustrate in a dynamic fashion the impact of the laser on red cells, showing the significance of their color and qualities. They provide dramatic movies for showing the dynamics of laser radiation in living tissue.

TISSUE CULTURE STUDIES

The studies of Malt and of Rounds and his associates with tissue cultures have expanded the field of laser microscopy. These studies show clearly the increased effect of the laser on those cells which have pigment such as retinal epithelium and melanoma. They show also the disturbed function of the cells such as heart muscle cells after laser impacts and the effects of the addition of dyes, etc. In addition, Rounds and his group have worked with tissue cultures as a study of enzyme systems such as thymidine and adenosine triphosphate. These investigators provide, then, a dynamic background on the intimate details of the effect of the laser on cellular cytology.

Rounds believes that laser radiation within the finite limits of energy densities and power outputs is a problem of the wavelength absorption and that such studies on absorption qualities may be able to be used laser for the clinical applications of the laser. Rounds has also done preliminary studies on the toxicity of an extract of the impact site. Controls are needed in these experiments.

The recent studies of Amy and Storb on the selective absorption of the mitochondria of dilute solutions of Janus green B are of interest with radiation of these cells with laser beams of 6 microns and precise microsurgical techniques which can be done using laser microscopy.

A further development of laser microscopy is the use of this instrumentation to study the synergistic effect of laser radiation with other modalities of radiation such as x-ray or ultraviolet and the use of chemotherapeutic agents such as nitrogen mustards, methotrexate and 5-FU. With good controls laser radiation is given to tissue cultures treated also with other modalities of radiation and various cytotoxic agents. Rounds has shown that the laser may add to the effect of gamma radiation. As yet, similar experiments have not been done with grenz radiation, ultraviolet light, xenon light and the like.

These tissue culture experiments can be carried over into investigative studies in cancer research. Protein binding and cytotoxic dyes are also other fields of research for studies in laser synergism. These studies on pigments influencing the absorption of the laser beam are important for clinical applications of the laser.

CYTOGENETIC STUDIES

One other concern of the laser, since it is a form of radiation, is its cytogenetic property. This can be studied also with tissue culture techniques by the observation of the direct effects of laser impacts on mitosis and by cultures of tissues such as the skin after laser impact. In addition, the laser, with suitable controls, may be used on embryos, ova, intra-

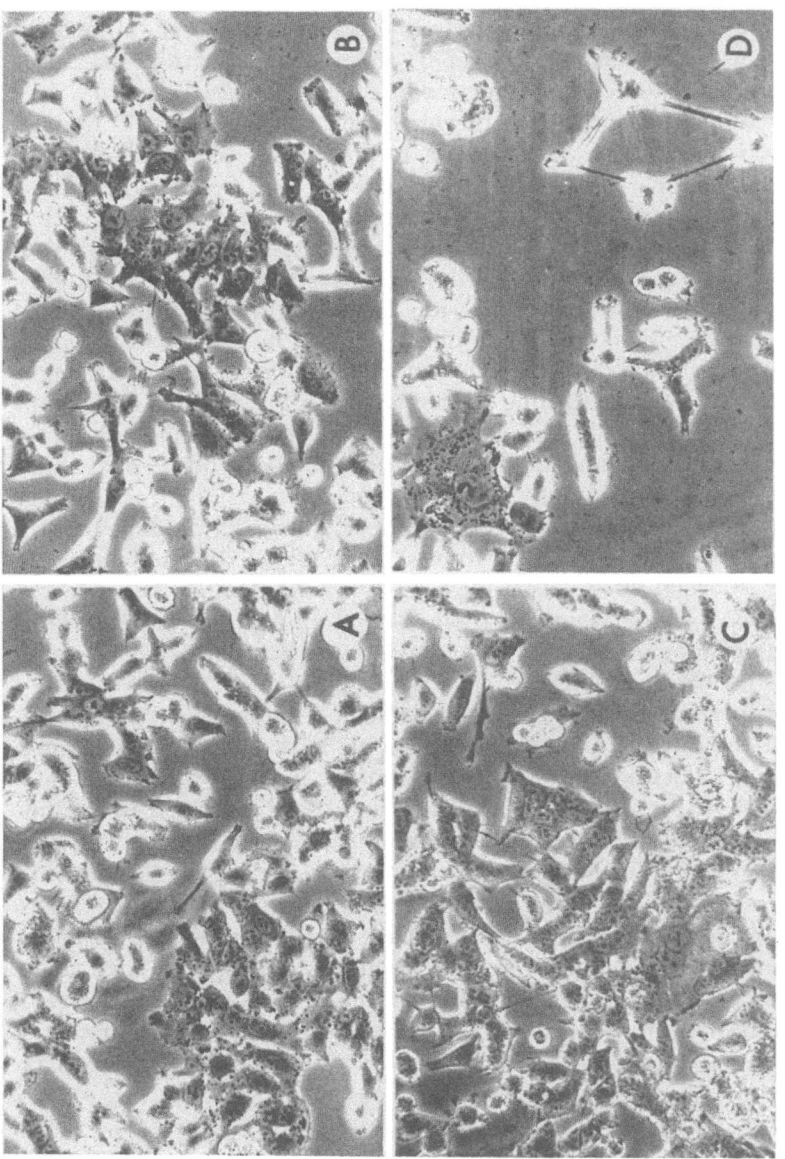

Fig. 11-3. Tissue culture technique to show qualitative effect of laser and gamma radiation. (A) Untreated adenocarcinoma cells *in vitro*; (B) Cells 4 days after exposure to 10 impacts of 2 joules each from normal mode pulsed ruby laser; (C) Cells 4 days after receiving 200r gamma radiation CO^{60}; (D) Cells 4 days after combined laser and gamma radiation. (*Reproduced with permission of Dr. Donald E. Rounds, Pasadena Foundation for Medical Research.*)

uterine radiation, and even on the testes. In his extensive tissue culture studies, Rounds has described relatively few cytogenetic changes. However, in one instance, the cultures of endothelial cells after laser impact showed abnormal chromosomes in several passages. Under the direction of Soukup, we have attempted to repeat this in our laboratory and also have cultured irradiated peripheral white cells. In no instance were chromosomal aberrations such as dicentrics and chromatid breaks found.

Sobkowski has recently developed a technique for selective retrieval and processing for chromosomal analysis of a pure epithelial cell population from the mucous membrane lining the Chinese hamster cheek pouch. The advantage over skin irradiation experiments is that the mucous membrane epidermis is thin, allowing for adequate measurement of radiation doses, especially for control studies with x-ray and Grenz ray for laser radiation. The pouch area may be easily manipulated for selective radiation and there are no adnexal structures. Although the Chinese hamster, according to Soukup, is difficult to handle and breed, it has only 22 chromosomes. We have adopted this technique of Sobkowski's for our laser radiation studies and as yet no chromosomal aberrations have been found.

In a subject in our laboratory who had impacted himself for 18 months with both ruby and neodymium lasers, blood chromosome analysis showed normal 46/XY karyotype by Warkany. Similar negative results in blood chromosome analysis were found in a patient with metastatic melanoma whom we had treated extensively with high energy ruby and neodymium lasers.

With proper controls, studies in experimental embryology can also be done with the laser. Under the direction of Takacs, experiments were done in our laboratory on selective radiation of embryos on one side of the rat uterus. As Edlow and his associates have described in similar experiments, large areas or radiation produced severe damage. In our studies using the Q-switched laser, in two of the rats of a series of four, radiated embryos died as well as one control. In one rat, the surviving embryo was lighter in weight. Hemorrhage in the rats which died was found about the cephalad area. Barnes and his associates have done fascinating studies in experimental embryology with the chick embryo and the rabbit ovum kept alive in culture medium. The laser has provided a precise microsurgical technique for studying specific effects in this field.

CONCLUSIONS

Basic studies in cytology and cytogenetics have already demonstrated that the laser beam can be used for microsurgical techniques in this fascinating field. This applies to laser radiation of tissue cultures of skin and mucous membrane and in experimental embryology with laser radiation of chick embryos and rabbit ova. New studies to be done should include selective radiation of human chromosomes and electron microscopic pictures of these immediate and subsequent changes as suggested by Hoskins. In teratologic radiation studies, studies on laser impacts of ova, intra-uterine, exposed testes and embryos of experimental animals should be continued with suitable controls.

Table 1

BLOOD CHROMOSOME ANALYSIS AFTER LASER RADIATION

59 year old Caucasian male   over 400 impacts
1½ year   left forearm   52 joules/cm² pulsed ruby laser

—	44	45	46	47	48	Total
—	—	4	36	—	—	40

40 cells counted
3 karyotypes made
Conclusion—46/XY karyotype

H. Maimon and J. Warkany, M.D., Children's Hospital Research Foundation, Cincinnati, Ohio.

Table 2

BLOOD CHROMOSOME ANALYSIS AFTER LASER RADIATION

30 year old Caucasian female   metastatic melanoma
multiple impact high energy ruby and neodymium   laser
70–1160 joules exit energies 48 hours after laser radiation

—	44	45	46	47	48	Total
—	4	6	29	—	—	39

3 karyotypes made:
1–46/XX apparently normal
1–45/XX with one in the G group missing
1–45/XX with one #16 missing

—	44	45	46	47	48	Total
—	1	5	32	—	1	39

4 karyotypes made:
4–46/XX apparently normal

L. Smith and J. Warkany, M.D., Children's Hospital Research Foundation, Cincinnati, Ohio.

Table 3

SKIN CHROMOSOME ANALYSIS AFTER LASER RADIATION

30-year old Caucasian female   metastatic melanoma.

Multiple laser impacts from ruby and neodymium lasers on melanoma and normal skin   70–1160 joules exit energies.

48 hours after laser radiation.

43	44	45	46	47	48	Total
1	2	3	52	2	—	60

60 cells counted

6 karyotypes made:

    4–46/XX apparently normal chromosomal constitution

    1–47/XX with an extra No. 2 chromosome

    1–47/XX with an extra No. 8 chromosome

Summary: Normal 46/XX karyotype in cells (fibroblasts) cultured from skin treated with laser.

J. Warkany, M.D. and M. Nieman, Children's Hospital Research Foundation, Cincinnati, Ohio.

Table 4

LASER MICROSCOPY

Goals

    1. Mechanism of laser action in cells and tissues

    2. Microsurgical manipulations

    3. As microprobe for spectroscopy

Instrumentation for Laser Attachment

    Ruby—output 70 microjoules to 700 millijoules

            focus spot 2–5 microns

            energy densities on tissue 10^3–10^7 joules/cm^2

            pulse length 500 microseconds

    Other lasers—attachment couplings as external sources

    Focus—3 mms. above stage

    Phase contrast lenses

    Time-lapse photography

    Laser microprobe—Q-switched ruby laser

            50 micron target spot

            spark-gap for further excitation

    Laser holography

Laser Safety

    Metal plates

    Filters

    Remote circuit TV

REFERENCES

Amy, R. L. and R. Storb: Selective mitochondrial damage by a ruby laser microbeam: An electron microscopic study. *Science,* 150:756, 1965.

Barnes, F. S., J. C. Daniel, and Takahashi: Laser surgery and biological damage. Presented at NEREM, Boston, 1965. Abstracted in *Laser Focus,* December 1965.

Bessis, M., et al.: Chemotactism after destruction of a cell by laser microbeams. *C.R. Soc. Biol. Paris,* 158:1195, 1964.

———— F. Gires, G. Mayer, et G. Nomarski: Irradiation des organites cellulaires a l'aide d'un laser a rubis. *C.R. Academie des Sciences,* 255:1010, 1962.

———— and M. Ter-Pogassian: Micropuncture of cells by means of a laser beam. *Ann. N. Y. Acad. Sci.,* 122:659, 1965.

Booth, et al.: Laser in cytology. *Nature,* 203:789, 1964.

Edlow, J., et al.: Laser irradiation: Effect on rat embryo and fetus in utero. *Life Sci.,* 4:615, 1965.

Goldman, L., D. J. Blaney, D. J. Kindel, III, D. Richfield, P. Owens, and E. Homans: Effect of the laser beam on the skin: Exposure of cytological preparations. *Jour. Invest Derm.,* 42:247, 1964.

———— and P. Owens: Chromosome studies in dermatology: Preliminary observations in some congenital and acquired dermatoses and in the effect of radiation with x-ray, Grenz ray and laser. *Acta Derm.-Venere.,* 44:268, 1964.

Hamberger, A., et al: The Q-switched laser as a tool in micro-diver technique. *Exp. Cell Res.,* 37:460, 1965.

Hoskins, G. C.: Electron microscopic observations of human chromosomes isolated by microsurgery. *Nature,* 207:1215, 1965.

Johnson, F. M., et al.: Effects of high-power green laser radiation on cells in tissue culture. *Nature,* 205:721, 1965.

Lang, K. R., F. S. Barnes, J. C. Daniel, and J. C. Maisel: Lasers as tools for embryology and cytology. *Nature,* 201:675, 1964.

Malt, R.: Effects of laser radiation on subcellular components. *Fed. Proc.,* 24:122, 1965.

Rounds, D. E.: Effects of laser radiation on cell cultures. *Fed. Proc.,* 24:116, 1965.

———— E. C. Chamberlin and T. Okigaki: Laser radiation of tissue cultures. *Ann. N. Y. Acad. Sci.,* 122:713, 1965.

———— R. S. Olson and F. M. Johnson: The effect of the laser on cellular respiration. Presented at NEREM, Boston, 1965.

Saks, N., C. Roth, and A. Charles: Ruby laser as a microsurgical instrument. *Science,* 141:46, 1963.

———— R. C. Zuzolo and M. J. Kopac: Microsurgery of living cells by ruby laser irradiation. *Ann. N. Y. Acad. Sci.,* 122:695, 1965.

Sobkowski, F. J.: Personal communication.

————: A technique for studying the in situ karyotype of epithelial cells from the cheek pouch of the Chinese hamster. *Exp. Cell Res.,* 33

Soukup, S. W., E. Takacs, and J. Warkany: Chromosome changes in rat embryos following x-irradiation. *Cytogenetics,* 4:130, 1965.

————: Personal communication.

Tomberg, V. T.: Biophysical effects of laser radiation. *Engineering in Medicine
and Biology.* Proceedings 19th Annual Conference, 1966, p. 59.

Warren, Shields and Loraine Meissner: Chromosomal changes in leukocytes of
patients receiving irradiation therapy. *J.A.M.A.* **193**:351, 1965.

12

Laser Effects on the Eye

Of all the viscera in the body, the eye is the most important one for all phases of laser research. It is the eye for which laser surgery can be done now. Also, it is the protection of the eye which is so important for all personnel engaged in laser research. The comparison is often made of the early stages of biomedical investigations of the laser to the early stages of the biomedical applications of x-ray. One does not want to repeat the tragedies of the early days of x-ray investigation. It has been said often that if one does not know history, one is forced to repeat all the mistakes which have occurred previously. Related to the eye hazards from the laser are the studies of the eye hazards of microwaves of radar. These are also of great interest and importance. Many of these studies on microwave effects on the eye are more detailed and advanced than those on the laser. The threshold value for the radar eye injury is given usually as 10 mw/cm², primarily with lenticular changes.

PROTECTION

Physicists are notorious for their disregard of protection in radiation work. It is difficult for one to appreciate the reactions which may occur from something which does not hurt and which often is not apparent, such as laser radiation with frequencies beyond the visible range. Moreover, if damage occurs about the periphery of the fundus, the lesions may be asymptomatic. Protection is often cumbersome, time-consuming and some programs may be even partially or wholly ineffective. As Rathkey has reported, serious accidents have already occurred to physicists. Yet physicists must realize in all phases of laser work that it is absolutely necessary for the eye to be protected. This includes even work with the low output gas lasers as well as with high energy lasers. With the recent introduction of high output gas lasers, such as the argon, protection of the eyes is difficult, but very important.

91

The details of personnel and area protection, and the details of the examination of all personnel working with lasers have been described in the chapter on area and personnel protection. It is up to the physicist and all people in and about laser facilities to use constantly protection in all phases of laser research. The physicists must help also with the development of laser protective devices, such as eye dosimeters or actual shields for protection of the eye. New lasers demand different measures for eye protection. For example, protective glasses for ruby and neodymium laser impacts will not necessarily protect against ultraviolet lasers or argon lasers. It is important to emphasize that although much is known about acute exposure, there is little known as yet of the effect in man of chronic exposure of the eye.

EYE STRUCTURES

The reason for the susceptibility of the eye to laser radiation lies in the fact that the lens system can focus laser beams and so increase the energy density in that area of the eye, the pigmented retinal epithelium, which can absorb laser radiation so easily.

Too little is known as yet about the changes in the so-called transparent tissues with transmission and reflectance of the laser beam and even some absorption of the incident beam of the laser. These structures are the aqueous humor and vitreous humor of the eye. The glistening outer layer of the eye is called the cornea. Laser beams may impact this surface as parallel beams to be focussed by the lens, or they can be focussed on this surface for investigative studies. How much of the beam is reflected from this corneal surface is not known. In Maxwellian view, the laser beam is focussed within the lens. As indicated previously, changes in white cells from laser impact may not be significant, but there is reduction of mobility following absorption of low energy laser beams. Low energy ruby lasers with 0.1 to 0.3 joule exit energy have not caused cataracts, and even after prolonged periods of observation none have been seen. Recently, Harger (personal communication) has indicated that repeated exposures (factors ?) by neodymium and ruby lasers may have been associated with rosette type of cataracts in two research workers. Geereats believes such lenticular changes can occur by repeated exposures to neodymium or if optical devices used could have caused the laser beam to be brought to a focus within or close to the lens. High energy laser impacts can damage not only the lens of the eye, but all structures. Ruby lasers have their greatest effect on the retina and choroid; neodymium on the ocular media, while the carbon dioxide laser affects the cornea and anterior segment. A number of studies have been done with the enucleated eye, but there the chief interest has been in the passage of laser light through the eye, and the detection with barium titanate transducers of ultrasonic shock waves produced by laser impacts. A crude proof of the shock wave produced in the eye on impact is to observe the enucleated eye as it is dashed from the holder or carrier upon impact of the laser beam. This is noted especially with the use of high energy lasers of 50 joules or more exit energy.

Any blood vessels in the sclera which have been impacted by the laser may develop hemorrhage or thrombosis according to the energy density of the beam.

As the laser beam is transmitted through the lens system of the eye, the beams are of wave lengths within or near the visible range, focussed by the lens system of the eye on the retina or pigmented part of the eye. Here, changes may be produced according to the type of laser, the exit energy, size of the target area, the duration of the pulse, collimation of the beam and also whether optical devices had been used. The damage extends beyond the measured target area of the incident beam through the absorption of the reflection of the laser. There is little information available as to the effect of the beam reflected from the pigmented retinal area onto other adjacent structures of the eye. At times, there may be no gross lesion at all, at least from low energy lasers. Microscopic sections including histochemical studies, or electron microscopy sections, enzymatic studies and fluorescein studies in the living eye however, may reveal these early lesions.

One of the difficulties of the control of the early lesion by the microscopic section is the inability often to find the impact site on sectioning the eye. The microscopic section shows different degrees of tissue destruction often with the excessive production of pigment. This is melanin pigment; occasionally, some blood pigment may also be found. Following impact, there is often increased production of pigment. Perforation of the eye may result also. Wolbarsht, Fligsten and Hayes, from studies of chorioretinal lesions in monkeys, believe most of the energy with the ruby laser is absorbed in the pigmented retinal epithelium. For detailed descriptions of the histopathologic lesions which develop immediately and at intervals following impact of the laser, one is referred to current work on laser eye irradiation by Zaret, Grosof, Ham and Geeraets, Zweng, Flocks and Martens.

Hemorrhage may develop in the eye after laser impacts. This hemorrhage may persist; it may also be absorbed gradually after a period of time. Actually the laser may be used also to "vaporize" some of the persistent hemorrhage remaining in the eye.

As indicated, with high energy impacts from the ruby laser, considerable destruction of even the transparent tissues of the eyes in primates may be produced, according to the studies of Jones and his associates. The need for such studies is to have some reference as regards accidents

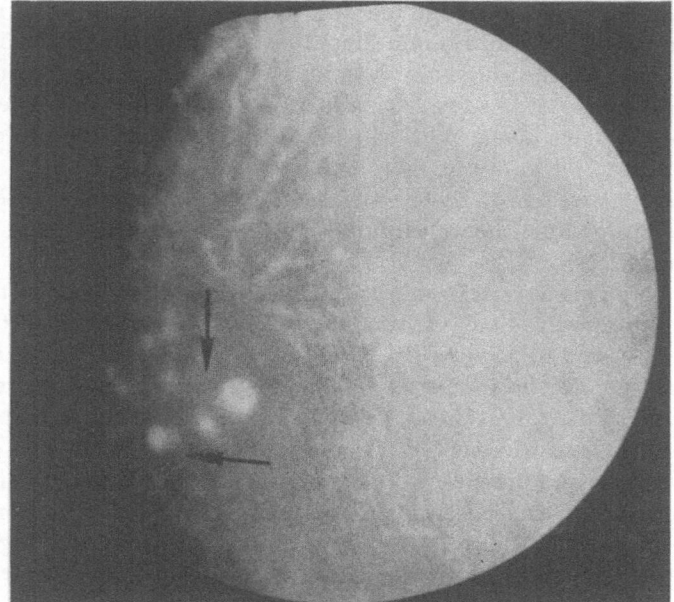

Fig. 12–1. Rabbit fundus, ruby laser, showing effect of varying energy densities, 0.1 to 0.3 joules. Black-and-white copy from Kodachrome (*Eastman*) picture.

to the eye and also for the studies of high energy laser in the treatment of cancer of the eye.

With high energy lasers, considerable burns may be produced even on the skin and hairs about the eyes. Low energy lasers may char the eyelashes. In rabbits, we have penetrated through the skull about the eye into the brain with impacts from high energy lasers, and caused spinal fluid to drip through the nostril on that side.

Ham believes that, unlike the CW and multiple spike operation lasers, Q-switched operation may exceed power densities of many megawatts/cm². He indicates that at such power densities and exposure times, it is doubtful whether temperature is a "meaningful concept in the sense of classical thermodynamics."

Investigative studies of laser transillumination of the soft tissues have been done about the face. No changes have been evident in the eyes of patients who have been treated with laser impacts for angiomas and cancers of the face. Long-term observation, however, will be necessary in this connection. The use of laser transillumination to detect masses, foreign bodies, bone defects, etc., is under investigation with special references to its value and its safety.

In addition to gross observation of eye changes and analysis of microscopic sections, now including electron microscopy, the functional

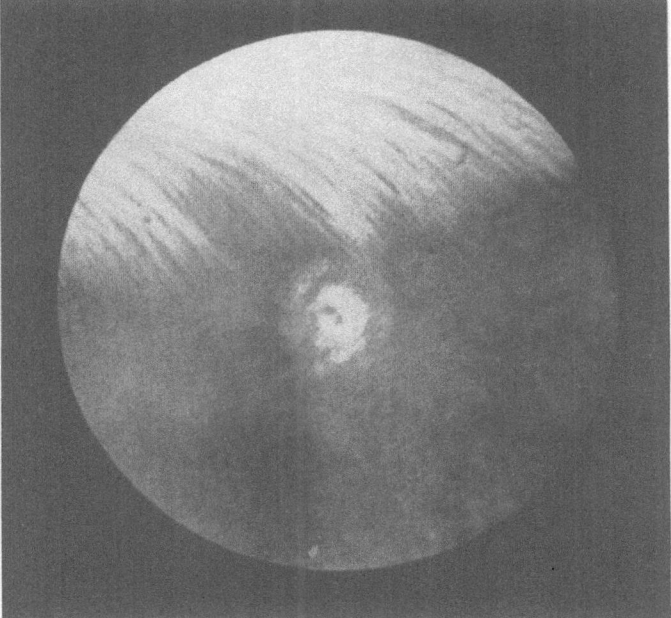

Fig. 12–2. Rabbit fundus, continuous wave argon laser (*Bell Telephone Laboratories*) 100 milliwatts, one second, four weeks after impact.

disturbances of color vision following laser impacts may be studied by the technique of electroretinograms. As yet, the electroretinogram is not of practical value in the detection of the so-called early or minimal laser lesions of the eye. Obviously, the early detection of minimal exposure is certainly the ideal for laser eye protection studies. Here, repeated examinations by the ophthalmologist and repeated fundus pictures of the macular area are required.

This is but a brief review of the essential basic effects of the laser on the eye. For detailed studies of spot size in the retina, threshold doses, intra-ocular temperature measurements and similar necessary basic studies on the eye after laser impact, one is referred to the reports of Zaret, Ham, Geeraets, Koester, and Campbell. As mentioned previously, in every phase of laser research the effects are influenced by the various parameters of the laser as regards the type of laser, whether ruby or neodymium, exit energy, target area, whether focussed directly on the cornea or impacted with parallel rays, duration of the pulse, and character of the pulse. The non-linear, non-thermal effects of the laser are of great significance, especially for the eye. Here, the cavitation effects of laser impacts with the production of ultrasonic waves as shown by Amar are of importance. Cytogenetic changes in retinal epithelium after laser

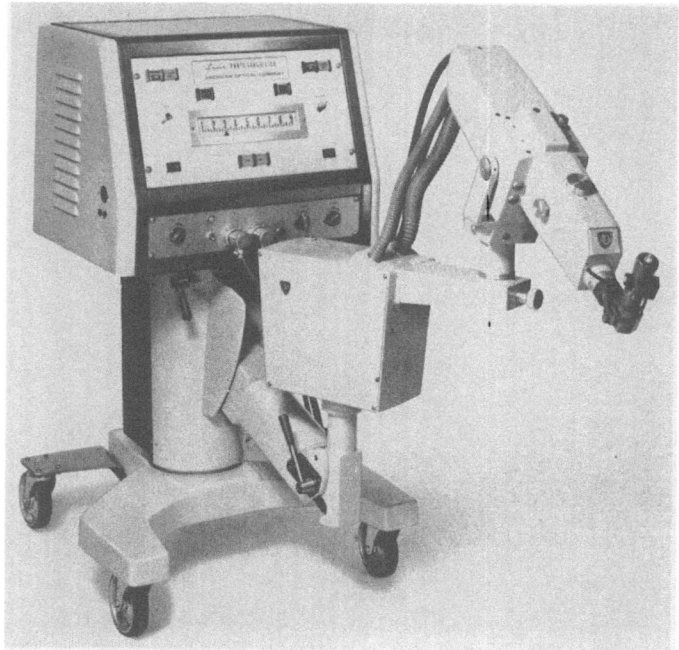

Fig. 12–3. Laser photocoagulator *(American Optical Co.)* .

impact will have to await long-term observation studies of the eye. Pigmented retinal epithelium has been used for tissue culture studies by laser radiation although it is difficult to maintain successive generations of cells.

LASER EYE INSTRUMENTATION

The basic purpose of laser eye instrumentation is to bring to the eye a laser beam which can be directed to the exact area desired with a very short exposure time and with minimal absorption by ocular tissues other than the target area. This means accurate control also of the size of the beam, duration of the pulse and exit energy. Control of the output is especially necessary for repeated impacts. The instrument should be flexible enough to have its laser beam reach all areas of the fundus. In addition, there should be protection for the operator. Unfortunately, many instruments now available do not satisfy these few basic criteria. The laser ophthalmologist must be one who has worked with laser radiation and understands this form of radiation and the values and limitations of laser eye instrumentation as well as the special fields of ophthalmology for which laser surgery is needed. A laser is not the type of instrument which the practicing ophthalmologist can pick up at a medical meeting, be impressed by the advertising prospectus and promptly start to use on his patients. A long apprenticeship then is necessary in the laser laboratory, followed by investigations on animals before treatment of patients is done.

In brief, the instruments available today are the modified hand laser eye instruments such as those of Kapany of Optics Technology, Inc., and of Honeywell, Inc. and the retinal coagulators of American Optical Company and Maser Optics, Inc. At present, the Maser Optics retinal coagulator, as it is, is suitable only for animal experimental work. Following the detailed investigations of Campbell, at present the American Optical laser retinal coagulator instrument is preferred. With handheld lasers, accurate control is necessary. Some operators prefer foot controls to the use of hand control buttons. With the portable coagulators, air cooling must be adequate to insure constancy of exit energy. Actually, however, unless each output is monitored, one is forced to judge impacts by the intensity of the retinal lesion produced. Outputs of retinal coagulators vary from less than 0.1 joule to 0.7 joule. Campbell and his associates have developed a neodymium fiber laser for the production of sharply localized chorioretinal lesions. They believe this to have value as an adjunct to photocoagulation and as a possible replacement for electro-surgical devices in retinal surgery.

LASER TREATMENT

After long experience with animal experimentation, especially with rabbits and, where possible, collie dogs with retinal detachments, the ophthalmologist may be ready to start the treatment of patients. Even then, it would be better to initiate experiments with eyes which may have to be enucleated later for various conditions. Even though these are not normal eyes, the operator learns the technique and the energy densities required for the particular color of the eye and for the specific lesion. The clinical studies of Campbell, Zweng and Flocks and Martens should be reviewed in detail, as well as reading the important and detailed studies on laser eye reactions by Zaret, Grosof, Ham, Geeraets, Schepens and Tengroth.

In brief, the eye conditions today for which laser surgery is used include retinal tears, flat detachments and angiomas and tumors. Some current investigative studies are the use of the laser in diabetic retinopathy to attempt to prevent progressive vascular damage by thrombosing blood vessels, for the treatment of uveitis, central serous retinopathy and edema of the macula.

The rabbit eye is used most frequently for animal experimentation. Eyes of primates are studied also. For dogs, it is usually that breed of collie with retinal detachment which is used. Cat's eyes may be used also. In the rabbit eye, after laser impacts, small circular whitish areas appear, followed by hyperpigmented areas and complete healing and scarring or perforation, depending on the energy density used. Control studies of all animal experiments should include fundus pictures and microscopic sections, including those with electron microscopy, and, where possible, electroretinogram tracings. As the color of the fundus varies from individual to individual and from animal to animal, the operator must take this into account in determining the dose of the first impacts. Once the impact is delivered, it is too late to recall it so the first impacts are usually calculated below the proposed dose.

There are many confusions as to the values of the critical retinal energy density of the rabbit eye for the production of a lesion. This subject will be discussed also in the chapter on protection. The figure often given is 0.7 joules/cm², 0.8–1.2mm diameter retinal spot with a pulsed ruby laser, 175μsec. It is evident however that less energy is required at high peak power with Q-switching. Campbell, Rittler, Noyori, Swope and Koester indicate that the lowest threshold for both rabbits and humans is evident 24 to 48 hours after exposure. They found in one human subject a lower threshold in the macular region by a factor of 2.5x. The rabbit has a lower threshold than man.

Even though there is much need and desire for threshold figures on eye safety, accurate data are not available as yet. It is hoped that the laser eye protection committee of the National Research Council will soon give some standard figures for such threshold values for man, not only for normal and Q-switched ruby lasers, but also for neodymium and argon, carbon dioxide and nitrogen lasers. As yet, unfortunately, the data on eye damage from high output argon lasers, now rapidly developing from 1 watt to 200 watts, with their blue-green light, are not available.

The operator trains himself by the use of varying sizes of target areas and varying energy densities, as well as the ability to localize the laser impact where it is needed. The instruments should be capable of being manipulated freely, yet held rigidly when and where necessary to impact the retina.

For patients, the head must be fixed so that there is absolutely no movement. This ordinarily means that the patient should lie flat and the operator must work above him. Ophthalmologists may use a type of ophthalmoscope for direct or indirect ophthalmoscopy, depending upon their skill and their training. As indicated previously, at present, to most ophthalmologists using lasers, the American Optical Company laser retinal coagulator appears to be preferred for clinical use.

In human experimental investigative studies, the condition treated most often with lasers has been the retinal detachment. Here an effort

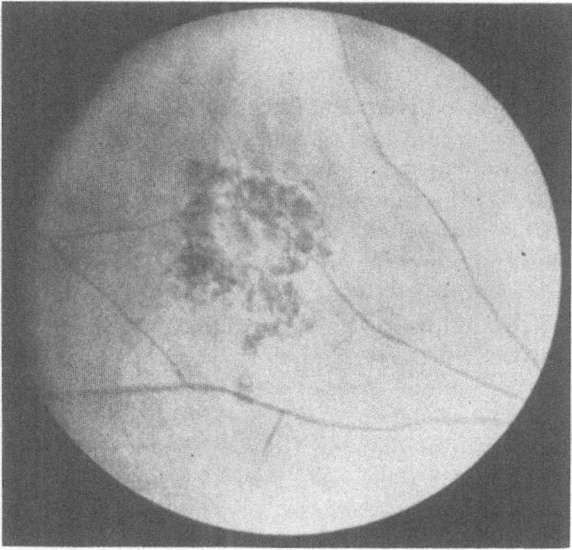

Fig. 12–4A. Retinal detachment—immediately after laser therapy. *Photograph by C. J. Campbell.*)

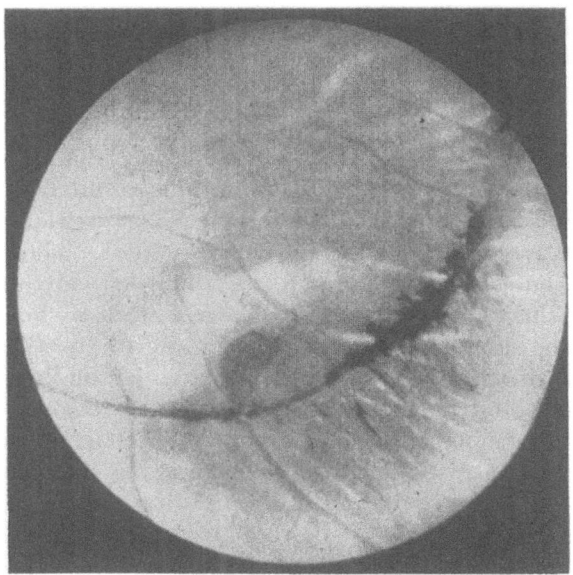

Fig. 12–4B. One month after laser treatment. (*Photograph by C. J. Campbell.*)

is made, with small size, low energy impacts about the detachment, to produce adhesion. These must be calculated accurately or otherwise additional hemorrhage, perforation or even increase of the detachment may occur. Impacts are given usually in the form of a circle of applications about the lesion. No hemorrhage is produced with the proper energy densities. As indicated, follow-up studies are necessary for prolonged periods of time. Patients complain of no pain but occasionally of warmth (?). They do see the brilliant flash.

Another condition which has been treated by those interested in laser ophthalmology is angioma of the eye. Here Campbell has felt it better to impact around the lesion rather than directly over the angioma. This method was utilized to prevent excessive hemorrhage. Reflection from laser impacts on angiomas increases exposure energies of other ocular structures, according to Geeraets. L'Esperance believes that retinal angiomas larger than two disc diameters in size do not respond to laser irradiation, partly because of their size. Retinoblastomas have been treated extensively by Campbell and others. The laser is available also for the impact of other tumors of the eye with low energy densities since pigmented tumors will show some selectivity to ruby laser radiation. This is not different from any white light source. This is true especially for those lesions near nerve fiber layers. Results of these treatments are not known as yet. Zweng and Flocks have done preliminary experiments on the use of the laser to try to destroy corneal scars to permit penetration of light into the eye. Martens has also considered the impact of the ciliary process to try to use that in the treatment of glaucoma.

Another phase of eye pathology which is receiving considerable attention at the present time is the treatment for the prevention of hemorrhage by the induction of thrombosis by laser radiation. This is of interest as regards the prevention of the changes which occur from the vascular reactions of diabetes. Root has mentioned that the laser is being used to treat diabetic retinopathy or retinitis proliferans. For this he mentions the laser is superior to the xenon photocoagulator. Fluorescein is also being used in diabetic retinopathy studies and in other eye conditions. Energy densities have to be controlled carefully; otherwise, hemorrhage can be produced by the laser instead of being prevented. One technique used by Martens in the study of the threshold values of energy densities for the production of vascular thrombosis in retinal vessels of the rabbit has been the use of dyes injected into the carotid artery and subsequently the impact of these vessels with a laser beam. Accurate focussing with small target areas on the vessel with low energy densities may help to induce spreading thrombosis. From his studies on the laser treatment of retinal microaneurysms, L'Esperance noted enlargement and engorgement immediately after laser treatment. Most of the lesions, but not all, were obliterated gradually.

The eye, then, offers considerable promise for contined investigations with the laser. The pioneering studies of Campbell, Zaret, Zweng and Flocks, Schepens and Ham and Geeraets will be the bases for any continued clinical studies of laser irradiation of the eye. The majority of laser eye studies have been done with the ruby laser. Clinical studies with the neodymium laser have just begun. The use of neodymium doped glass fibers for direct impacts in the eye has been investigated by Campbell.

There is still considerable doubt as to the true value of the laser co-agulator over the xenon light coagulator. Some of the confusion may be resolved with the development of increasing dependability and increasing flexibility of laser eye instrumentation, greater experience with efficient instrumentation and also by critical studies by experienced ophthalmologists who have tried both xenon photocoagulation and laser coagulation.

Those ophthalmologists who have had the most extensive experience with laser are more enthusiastic and more cautious than the practitioner who uses this occasionally. Campbell with his long experience is still enthusiastic. Geeraets believes that over-exposure to the laser is the greatest difficulty and consequent danger. For photocoagulation for retinoblastomas, Campbell believes the laser is good for prophylactic photocoagulation and is satisfactory for treating even the far periphery of the eye.

The follow-up of patients who have had laser surgery is of importance as regards the development of scars, recurrences, chronic change in the fundus, etc. As yet, hemorrhage has been the major reaction. No opacities of the lens have been reported in patients now followed for three years.

This clinical review, then, is to give the biologist, physicist and the engineer some background in the goals of laser surgery of the eye. The best way for them, if they are interested in laser eye studies, is to work with the laser ophthalmologist. This ophthalmologist, as has been stated repeatedly, also needs the laser physicist and the laser instrument engineer for his own studies. The ophthalmologist must be available at all times to the laser laboratory to give the pre-employment eye examination, to do ophthalmoscopic examination at intervals during the employment and to review constantly the laser eye protection program. Personnel, then, with the awareness and interest of the ophthalmologist, will have a proper respect for the continued need for protection against laser radiation.

FUTURE NEEDS

For future studies of the eye in relation to the laser, then, there is the ever-present need for instrumentation development and design. This will include the fundus camera attached to the proper laser eye coagulator and provided with a screen so that an enlarged view of the fundus, before, during and after laser impact is available. Accurate laser detection instrumentation should monitor the laser so that accurate measurements can be made, including micro-miniature intra-ocular techniques. The details of the eye protection program as regards the establishments of minimal exposure factors for all types of lasers, area control including monitoring instruments and personnel control with effective eye protection will help to insure laser safety for all concerned. Investigative studies should continue on active and chronic exposure of animal eyes with special emphasis on long-term observation of chronic exposure of animal eyes. There is need also for complete pictures of the dynamic response to the impact of different lasers for all structures of the eyes, not only the significant retinal epithelium. Histochemical studies, electron microscopy, electroretinograms and tissue cultures of impact and control areas will be needed to complete the picture. Electron spin resonance spectroscopy of laser impacted tissues, with suitable controls, should be compared to that of pigmented and non-pigmented eye tissues impacted by xenon light, ultraviolet light, infrared, ionizing radiation and electrosurgery. The greater use of intra-vascular fluorescein techniques will aid in these studies. Clinical studies by experienced clinical investigators should continue. As Zweng, Flocks and Peabody have done, more studies are needed of the histopathology of human ocular laser coagulation. Other clinical studies include the use of the helium-neon gas laser as illumination for the ophthalmoscope especially in fundus examination in patients with cataracts. The development of the argon laser as a retinal coagulator is also under study.

CONCLUSIONS

It is with such critical and detailed investigations that the future of laser eye surgery can be determined and also the future of the laser as an experimental tool for work in biophysics, physiology and biochemistry of the eye can be established. At present, eye treatments for patients should be limited to the medical centers where training, experience, interest, adequate equipment and research and follow-up facilities are available.

REFERENCES

Asayame, R.: Trial manufacture of a laser photocoagulator. Preliminary report. *J. Clin. Ophthal.* (Tokyo), 17:1369, 1963.
Campbell, C. J., et al.: Intraocular temperature changes produced by laser coagulation. *Acta Ophthal.,* Suppl., 76:22, 1963.

———— K. S. Noyori, M. C. Rittler, and C. Koester: Clinical use of the laser retinal coagulator. *Fed. Proc.,* Suppl. 14, S-71, 1965.

———— ———— ———— ————: Retinal coagulation clinical studies. *Ann. N. Y. Acad Sci.,* **122**:780, 1965.

———— ———— ———— R. E. Innis and C. J. Koester: The application of fiber laser techniques to retinal surgery. *Arch. Ophthal.,* **72**:850, 1964.

Campbell, C. J., M. C. Rittler, K. S. Noyori, C. H. Swope, and C. Koester: The threshold of the retina to damage by laser energy. *Arch. Ophth.,* **76**:437, 1966.

Dunnington, et al: Light coagulation, Ophthalmic surgery. *N. E. Jour. Med.,* **269**, 8.

Fine, S. and E. Klein: Biological effects of laser radiation. *Advances in Biological and Medical Physics,* John H. Lawrence and John W. Golfman, eds., New York, Academic Press, 1965, p. 147–212.

Freeman, H. M., O. Pomerantzeff, and C. L. Schepens: An evaluation of the ruby laser as a retinal coagulating source. *Ann. N. Y. Acad. Sci.,* **122**:783, 1965.

Geeraets, W., W. T. Ham, Jr., R. C. Williams, H. A. Mueller, J. Burkhardt, D. Guerry, III, and J. J. Vos: Laser versus light coagulator: A fundoscopic and histologic study of choreoretinal injury as a function of exposure time. *Fed. Proc.,* Suppl. 14, S-48, 1965.

Ham, W. T., Jr.: Effects of laser radiation on the mammalian eye. *N. Y. Acad. Sci.,* January 20, 1966.

Jones, A. E.: Effect of the laser on primate ocular structures. *Laser Focus,* **1**:6, July, 1965.

Kapany, N. S.: Retinal coagulation by lasers. *Nature,* **199**:146, 1963.

Koester, C. J.: Design and performance of a laser photocoagulator. *Laser Focus,* **1**:6, July, 1965.

———— Snitzer, C. J. Campell, and M. C. Rittler: An experimental laser retinal coagulator. Presented at the Optical Society of America, Washington, D.C., March, 1962.

———— M. R. Thornburn, H. Jupnik, and C. J. Campbell: Design and performance of a laser photocoagulator. Presented at NEREM, Boston, November 1965. Abstracted in NEREM Record, p. 154.

Kohtiao, A., I. Resnick, J. Newton, and H. Schwell: Temperature rise and photocoagulation of rabbit retinas exposed to the cw laser. *Amer. J. Ophthal.* **62**:524, 1966.

———— ———— ———— ————: Threshold lesions in rabbit retinas exposed to pulsed ruby laser radiation. *Amer. J. Ophthal.* **62**:664, 1966.

L'Esperance, F. A., Jr.: Laser treatment of retinal microaneurysms. *Medical World News.*

Martens, T.: Personal communication.

Pomerantzeff, O.: Comparison between laser and xenon light sources in photocoagulation, *Laser Focus,* **1**:6, July, 1965.

Rathkey, H. S.: Accidental burn of the macula. *Arch. Ophthal.,* **75**:346, 1965.

Root, H. F.: Personal communication.

Straub, H.: Protection of the human eye from laser radiation. Report from the Harry Diamond Laboratories, Army Materiel Command, July 10, 1963.

Swope, C. H. and C. J. Koester: Eye protection against laser. Report from the American Optical Company, Southbridge, Massachusetts.

Tengroth, B., B. Karlberg, and T. Berquist: Laser action on the human eye. *Acta Ophthal.*, 41:595, 1963.

Wolbarsht, M. L., K. E. Fligsten, and J. R. Hayes: Retina pathology of neodymium and ruby laser burns. *Science*, 150:1453, 1965.

Zaret, M.: Analysis of factors of laser radiation producing retinal damage. *Fed. Proc.*, Suppl. 14, S-62, 1965.

————: Laser hazards. *Laser Focus*, 1:3, January 1, 1965.

————: Ocular lesions produced by an optical maser. *Science*, November, 1961.

———— H. Ripps, I. M. Siegel, and G. Breinnen: Laser photocoagulation of the eye. *Arch. Ophthal.*, 69:97, 1963.

Zweng, H. C. and M. Flocks: Clinical experiences with laser photocoagulation. *Fed. Proc.*, Suppl., 14, S-65, 1965.

———— M. Flocks and R. Peabody: Histology of human ocular laser coagulation. *Arch. Ophthal.*, 76:11–15, 1966.

13

Laser Effects on Blood, Blood Vessels and Blood Vessel Tumors

Impacts of the laser on living tissue show significant effects on blood vessels with either hemorrhage or thrombosis (clotting), or both. It was assumed that the red color absorbed the laser radiation. Our studies on fresh blood smears showed destruction of red cells even with low energy densities. These changes included swelling of the cell, blister formation and even complete vaporization. The energy densities for such changes in red cells of fresh smears were found by J. Goldman and P. Hornby to be in the range of 0.16 joules exit energy. It was noted that white cells in the same field showed little effect when these blood smears were stained after exposure. There was little effect on platelets unless dyed.

BLOOD CELLS

To determine any differences in the laser destruction of different types of red cells, abnormal red cells were studied by Forristal in our laboratory. Fresh blood of normal, sickle, spherocytic types and red cells from newborns were examined. The impacts were done with the laser attached to the microscope and the energy densities required for destruction of the abnormal red cells and those from newborns were the same as for the normal red cells. Rounds, Olson and Johnson found that a power density of 100 MW/cm^2 of green laser radiation made human red cells transparent to wave length 400–600 mu. This was not possible with similar power density from a ruby laser.

Normal red cells were hemolyzed and the hemoglobin solution was exposed to both focussed and unfocussed beams of the ruby laser at low energy densities. The control and exposed samples were measured spectrophotometrically at wave lengths of 3500 Å to 7500 Å. Spectrophotometrically, hemoglobin, cyanmethhemaglobin and oxyhemoglobin were all unchanged. Similar negative results of exposure of hemoglobin have been found by Klein. Rounds, Olson and Johnson found that oxygenated hemoglobin after laser radiation of green (530 mu) laser showed no detectable effect. Yet, the same radiation changed the shape of the absorption spectrum of reduced hemoglobin to one that was intermediate between reduced and oxygenated curves. In addition, Forristal studied the globin of hemoglobin with starch block electrophoresis. In preliminary studies, there was migration of the laser irradiated hemoglobin in a double eardrop formation with the non-irradiated area migrating in a single drop formation.

White cells provide an excellent test material for the effect of the laser on so-called transparent tissues. However, even these cells, in a strict sense, are not completely transparent. Perhaps no living tissue truly is. The apparent negative results of the laser impact of the white cells in fresh and dried blood smears have been noted. These white cells, then, appear negative in the stained smear after laser impact. Forristal has studied the laser radiation of cultures of peripheral white cells. The white cells were exposed to 50 to 70 joules exit energy by unfocussed beam of the pulsed ruby laser. Peroxidase activity was increased over the controls. However, there was no change in alkaline phosphatase activity. White cell motility was generally depressed. Forristal noted also that in the segmented neutrophiles the nuclei were placed to the extreme rear of the cell wall and appeared to be in a fixed, non-pliable state after laser impact. Studies of the India ink technique of phagocytosis to determine effect of the laser impact on this property are still incomplete.

BLOOD PROTEINS

Serum albumin did not show any significant changes as compared to the controls. However, serum gamma globulin did show a significant reduction. Fine and his associates found that ruby irradiated human gamma globulin had more precipitative activity with antiserum and less with rheumatoid factor than control preparations. These effects are different, they believe, from these of thermal denaturation of gamma globulin. Such experiments should be expanded with the addition of dyes, and other radiation controls such as xenon light, ultraviolet, Grenz and x-rays.

As an approach to the study of the synergism of x-ray and laser, first described by Rounds, Forristal exposed three-day old white cell cultures to laser, to x-ray and to combined laser and x-ray. As demonstrated before, x-ray radiation destroyed many growing white cells as did the laser. However, when both laser and x-ray were used, the quantity of cells that continued to grow to seven days was much less than the quantity growing of non-irradiated cells, but there was no additive effect.

One other experiment with laser on blood elements was radiation of human plasminogen obtained through Mertz. After radiation this showed some changes as regards decreased optical density, decreased esterase activity and the development of a cloudy solution in distilled water as opposed to the clear solution of the control. Starch-gel electrophoresis, however, showed no apparent change in regard to the great migration of the modified protein following laser radiation. It is not possible to indicate the significance, as yet, of these studies on the radiation of human plasminogen.

BLOOD VESSELS

In an attempt to study the laser radiation of the red cells of the living animal, the femoral vein of the dog was cannulized and the circulating blood exposed to the pulsed ruby laser. Numerous studies were done, first by Forristal and by S. Goldman in our laboratory. It was possible with this technique in a dog to irradiate through a glass tube the red cells for periods of time up to eight hours. With baseline studies of counts and chromosomal cultures of peripheral blood, the following results were obtained by S. Goldman.

1. Decrease in total red cells
2. Decrease in hemoglobin
3. Blood chromosomal analysis—no growth as compared to normal karyotypes in the non-irradiated controls.

S. Goldman also did laser radiation with the pulsed ruby laser, 40 joules exit energy, over the sternum of guinea pigs after removal of the skin and fascia. No significant changes in red cells, hemoglobin or white cells were observed up to 25 days after impact. It was not possible to determine the adequacy of the radiation, especially because of the size of the target areas and the unfocused beam. No controls were done with x-ray.

The effect of the laser on the heart has not been studied in detail. In our laboratory, impact of unfocussed and focussed, 2 cms., ruby laser, 70 joules exit energy, showed only initial slowing of heart rate, but no ECG changes. Following the studies of Naprstek, Brown and S. Goldman have done studies on the exposed heart of the dog. With a curved quartz rod through the auricle, impacts were given about the AV node. Two hundred joules exit energy were transmitted through the rod with energy density of approximately 2,000 joules/cm². Localized deep hemorrhagic areas were produced. ECG patterns showed essentially no changes in PR internal of 0.08 seconds. The AV node was not impacted.

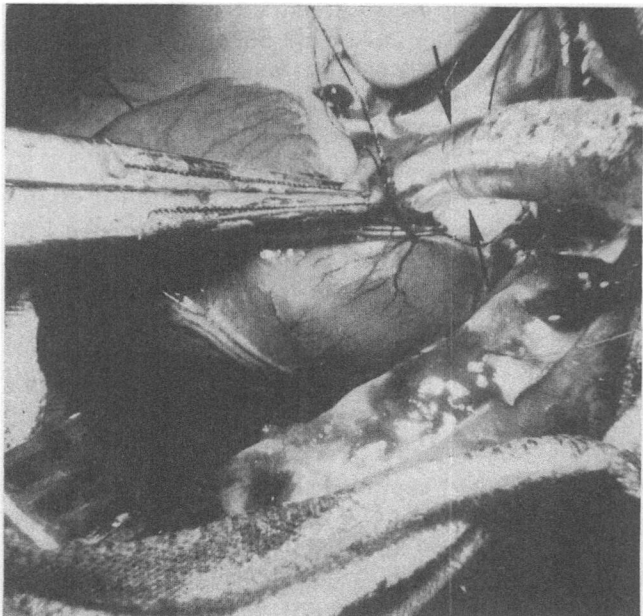

Fig. 13–1A. For transmission of pulsed ruby laser by a curved tapered quartz rod inserted through wall of auricle into a dog heart.

These experiments are being continued with controls, including thermal burns, electrocoagulation and xenon light. Complete heart block can be produced in this manner. The relative value of the laser over the other modalities must be determined.

Following the studies of the effects of the laser on various elements of the blood, the effects of the impacts on circulation were studied. Studies on man were done by observations of the laser impact of the capillaries about the fingernail bed and the laser impact of skin from the flexor surface of the forearm which had been stripped with Scotch tape. The impacts on the fingernail capillaries were observed more readily. Petechial hemorrhages and blanching of the capillaries about the target were observed. In an individual made reactive to the laser by repeated impacts, tiny white charred areas about the fingernails were produced by energy densities of 38 joules/cm² from a pulsed ruby laser. These energy densities did not produce reactions in the nail bed capillaries of previously exposed Caucasian volunteers. As yet, we have not done studies on measurement of blood flow after laser impacts with either thermistor sensors on the organ or intravascular bead thermistors.

Stripped areas of the forearm were difficult to visualize with capillaroscopy with the techniques employed by us. The excellent cinemato-

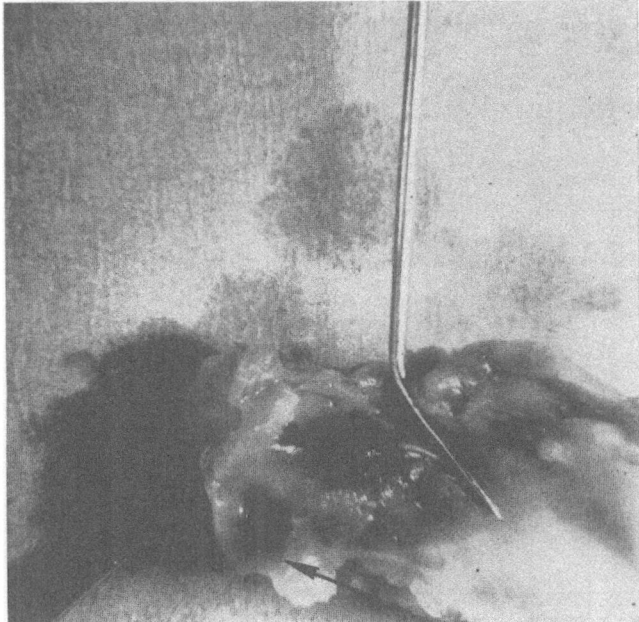

Fig. 13–1B. Showing localized area of impact from this quartz rod-transmitting, pulsed ruby laser with energy density 2,000 joules/cm². There was no reaction in wall of auricle about quartz rod.

graphic techniques of capillaroscopy by Demis were not used in these experiments. When blanching was obtained by the use of topical corticosteroid preparations, higher energy densities were required to produce change in surface impacts. Capillaroscopy pictures after topical fluocinolone acetonide showed significant whitish areas in the stripped skin. The vasoconstrictor effect, then, interferred with the absorption of the laser beam. We have also studied the effect of topically applied adrenalin, histamine and tetrahydrofurfuryl nicotinic acid ester on this stripped skin area after the use of the vasoconstrictive corticosteroid fluocinolone acetonide in various vehicles, including dimethyl sulfoxide (DMSO). Examination of the circulation of the conjunctiva under moderate magnification may be used of eye exposures to the laser.

In animals injected with dyes, the blood vessels absorbed the incident beam at lower energies. This has been shown by Martens with the injection of methylene blue in the carotid artery of the rabbit prior to laser impacts of the retinal vessels. A relatively simple technique for studying in a dynamic fashion the effect of laser on small blood vessels is to produce corneal scars in the eyes of rabbits and then to use blood vessels developing in the scars.

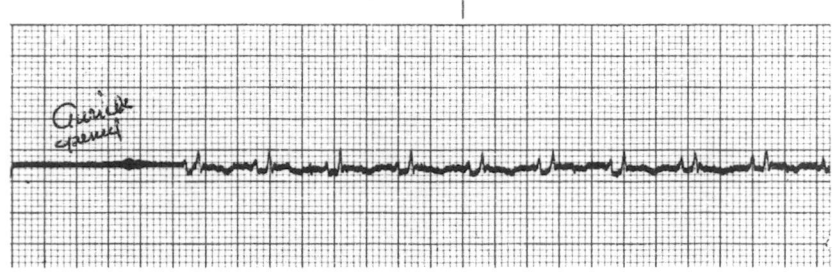

(1) Pre-impact; P R internal; 0.08 sec.

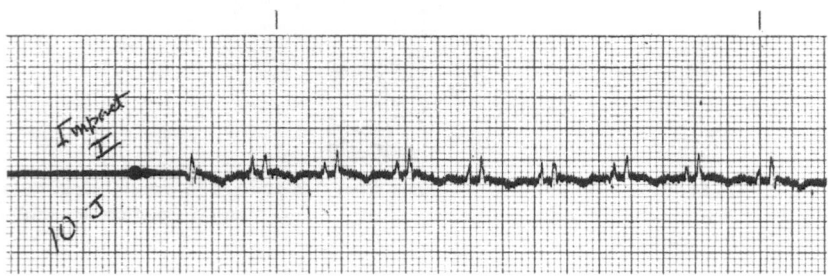

(2) Post-impact; 10 joules; P R unchanged 0.08 sec.

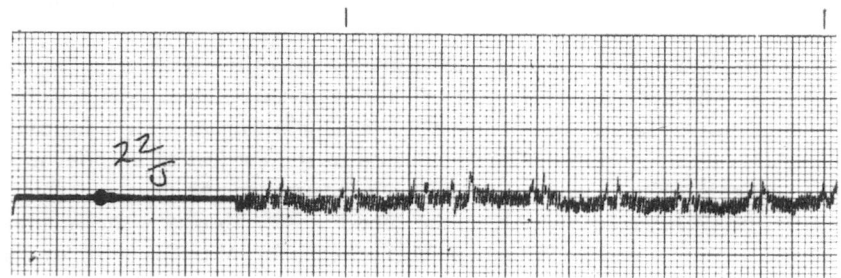

(3) Post-impact; 22 joules; with interference.

Fig. 13–1C. Electrocardiograms after series of impacts in dog heart; AV node not impacted (*D. Adair and S. Goldman*).

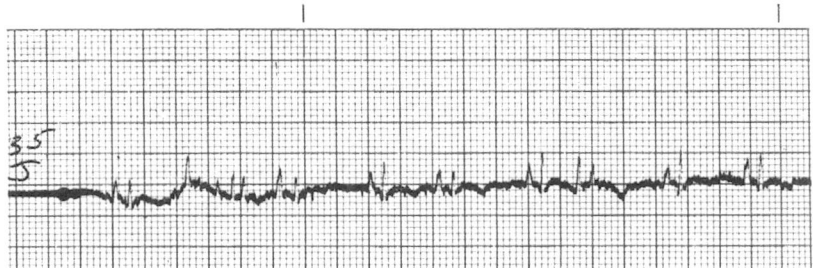

(4) Post-impact; 35 joules; P R 0.08 sec.

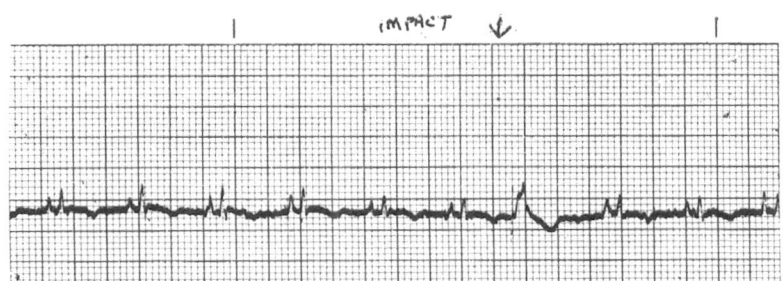

(5) 46 joules showing pre-impact; impact; and post-impact. No change.

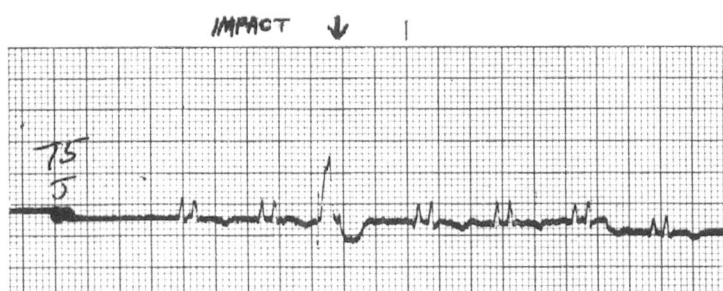

(6) 75 joules showing pre-impact; impact and post-impact. No change.

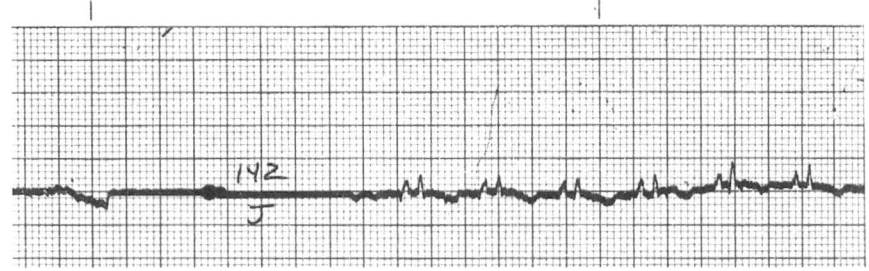

(7) Impact 142 joules; showing return to normal pattern. Unexplainable.

Another technique for the study of the effect of the laser on blood vessels has been the use of the meso-appendix preparation under the microscope as developed by Kochen and Baez. Dynamic pictures of the mechanisms of hemorrhage, thrombosis and recannulizations were obtained. It appeared briefly that energy densities made for the significant differences between hemorrhage and thrombosis. Increased energy densities caused hemorrhage rather than thrombosis. This had been noted also by Helmer in our laboratory in his study of impacts of the pial vessels of the spinal cord in dogs. His purpose was to develop selective localization of the laser impact for rhizotomy techniques for the relief of intractable pain. The irregular response from the laser made it difficult to attempt to produce necrosis of tissue without hemorrhage from the pial vessels. These features are discussed in more detail in the chapter on laser neurosurgery.

These studies of vascular dynamics are important, then, for laser impacts of all living tissues. Obviously, they are of significance in the eye where hemorrhage can develop and also in the brain where hemorrhage can develop and in the study of the laser impacts of internal viscera. In human skin small hemorrhages are often produced, especially by partially focussed or focussed impacts.

To induce thrombosis rather than hemorrhage more basic research will have to be done with control of such parameters as the blood itself, the type of laser, ruby, neodymium or argon, pulse duration and spiking, as well as variation in energy densities. Initial studies with the argon laser in areas like the liver indicate that there is more hemorrhage control with the argon laser. For these parameters of research, blood flow studies of engineers, such as those done by Cho and Hershey, will have to be used. Such factors as the blood elements, viscosity, before and after impact, blood vessel wall, blood vessel diameter, blood flow rate and mass transport through capillary walls are but a few of the phases which must be analyzed.

Lasers have also been used for the arrest of hemorrhage. This has been done in our Laboratory in arrest of bleeding from superficial angiomas. Klein and Fine have stopped hemorrhage with experiments in the excised tail area of heparinized animals. The control animals, without laser impacts, bled to death. Because of their effect on blood and blood vessels, argon lasers should be used in hemophiliacs for arrest of bleeding or for surgery. Preliminary studies with 10-watt output carbon dioxide laser indicate that this laser is less effective for controlling hemorrhage than the argon laser.

The laser has been used also by Yahr, Strully and Hurwitt to produce small vessel anastomosis. In this experimental technique, canine carotid arteries were used. A surgical adhesive was used to attach the donor vessel to the side of the recipient vessel. A drop of copper sulfate solution was used on the recipient vessel at the proposed site of the anastomosis. No selective stain is needed with the carbon dioxide laser because of the transparency of the blood vessel wall. Then another drop of copper sulfate was applied in the lumen of the common wall between the two arteries. The neodymium laser then produced a smooth hole in the common wall. The anastomosis may not hold because of the elastic recoil and pressure wave of the impact of the neodymium laser. Thus, the carbon dioxide and argon lasers may be more effective here. Studies in our laboratory have also been done on anastomosis by the use of quartz rods to provide a more selective localization of the areas of anastomosis.

ANGIOMAS

Extensive clinical studies have been done by us on angiomas of infants and children and the so-called "senile angiomas" of the adult. With ruby and/or neodymium lasers, small vessels are occluded rapidly. The development of coagulation necrosis often gives excellent cosmetic results as compared with those from the electric needle. Blanched areas have been produced in strawberry angiomas of infants with low-energy densities in the range of 40–50 joules/cm² transmitted through quartz rods. Efforts were made to study the transillumination of the skull in infants in these procedures, but no definite results of transillumination were observed with the infrared photographic technique. Low-energy impacts have been used also in angiomas in the mouth and gums.

Significant results were obtained with the use of laser for the so-called portwine angioma. This is a type of birthmark which has the appearance of wine spilled on the skin. There is no form of treatment for this and measures advocated, such as plastic surgery, often cannot be used because of the extensive development of these lesions. Tattooing to provide less color contrast is often not successful from a cosmetic standpoint. Ruby laser impacts at energy densities of 49–60 joules/cm² for the average reddish-bluish birthmark show significant lightening effects which may persist for a number of months. Test areas are done with both ruby and neodymium lasers. If the energy levels are kept to this energy density there is little danger of scarring. The treatment areas are, unfortunately, small, with the vast surfaces which these portwine angiomas cover especially those about the face. Many treatments are therefore necessary. General anesthesia has been used for extensive laser therapy. Cardboard has been used to protect adjacent areas of normal skin and other areas to avoid over-exposure. The cardboard can be cut to the peculiar design of the portwine angioma. Detailed histologic and histochemical studies show spotted effects not only on the blood vessels, but also on the dermis with spotty areas of fibrosis. Recent improvements of the radiation techniques have included the use of skin

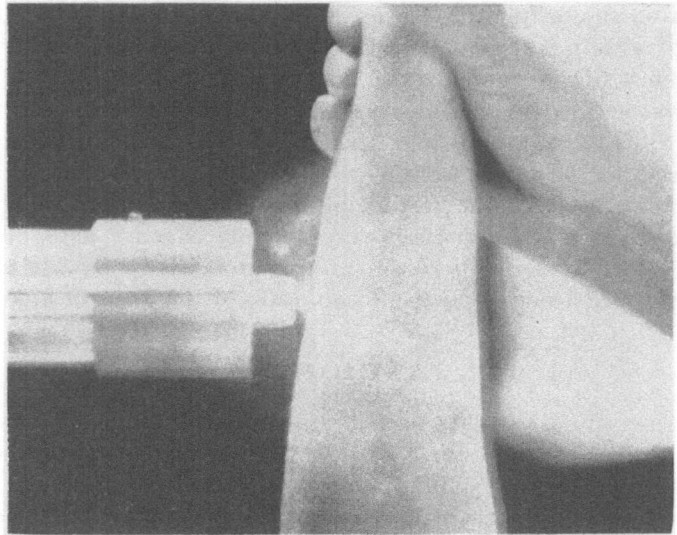

Fig. 13-2. Laser plume and bulging of the skin after laser impact on portwine angioma of forearm, pulsed ruby laser, 40 joules/cm². Black-and-white copy of Koda-chrome (*Eastman*) picture of impact.

stripping prior to impact to remove the protective and reflectant keratin layer and negative lenses with high energy ruby laser to produce target areas of 2–3 cms. in diameter, and the use of the argon laser. The general impression is that the neodymium laser of energy densities similar to that of the ruby gives less significant superficial charring reactions than the ruby, although the neodymium may produce a deeper response. The observation periods for these studies have been only three years and, consequently, more details are needed. However, there is enough knowledge at this time to indicate that trial test areas, so-called "postage stamp" size, may be used in one portion of the angioma. Then, the effects of the laser impact should be observed carefully over a period of several months before a decision is made as to the need for more extensive therapy. The incidence of revascularization is not yet known. Eye protective devices, similar to those used for x-ray therapy about the eyes, are under study for laser therapy about the eyelids. There is no evidence, as yet, that the laser treatment of these portwine cases have given rise to any Sturge-Weber reactions previously not evident to the patient. These are convulsions produced by brain involvement by the angioma about the face.

The laser has been used also in the treatment of malignant lesions developed from blood vessels such as Kaposi hemorrhagic sarcoma and true angiosarcoma. When the lesions are small, the clinical effect is excellent. With multiple or extensive lesions, as yet, laser therapy is not

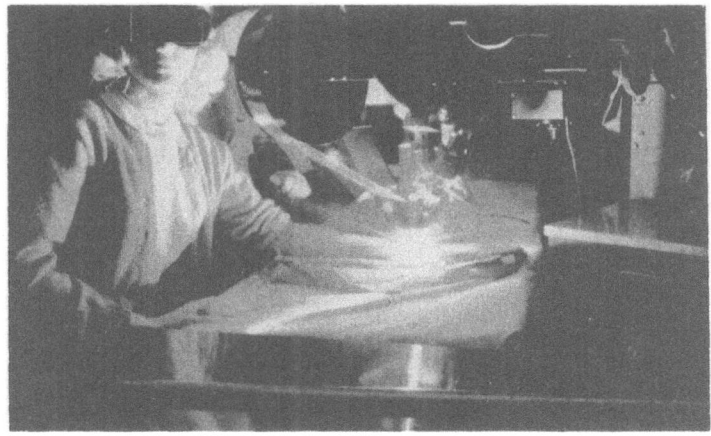

A

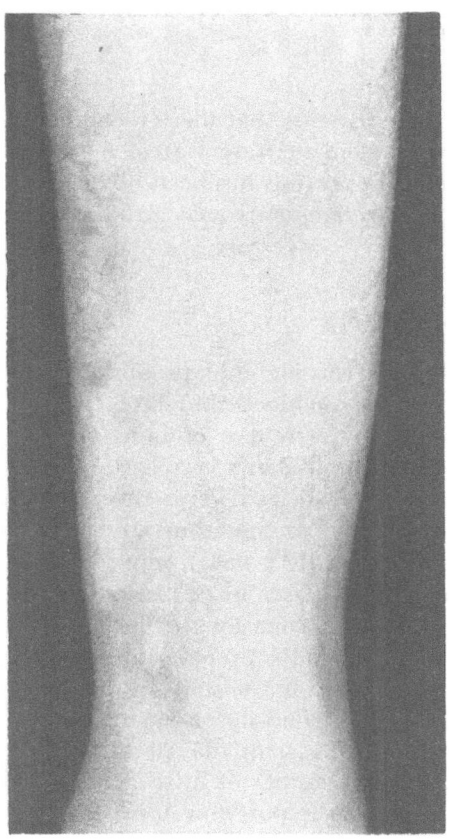

B

Fig. 13–3A. Showing laser impact on portwine angioma of forearm. Black-and-white copy of colored infrared Ektachrome (*Eastman*) picture of laser impact.

Fig. 13–3B. Showing effects after four months of repeated laser impacts on portwine angioma of forearm.

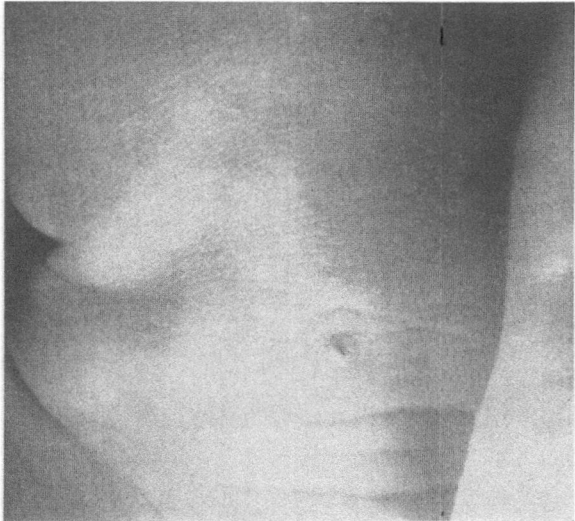

Fig. 13–4. Impact on angioma on heel of an infant (multiple eruptive type) immediately after impact 20 joules/cm², pulsed ruby laser.

practical. There was no evidence in these patients that there was progressive spread of laser coagulation necrosis from the target area. A similar lack of spread of laser-induced coagulation necrosis has been observed in preliminary studies of laser treatment of angiosarcomas produced by Swarm in mice.

CONCLUSIONS

It is evident, then, that the laser has considerable promise in the field of investigative and clinical research on blood and blood vessels. The membranes of the red cells may be the source of much needed study since after laser impact hemoglobin itself shows very little change. The effect of the laser on the blood vessel wall, as regards the intimate structure of these walls, is not known at the present time. In order to develop the laser for vascular surgery, especially arterial, more research will have to be done in this particular field. Laser instrumentation will continue to be the laser attached to the microscope for the effect on the blood and blood elements and for meso-appendix preparations and the laser with and without the use of quartz rods for transmission for observation on the effect on living blood vessels. The development of fiber optics permitting significant transmission of laser or the use of neodymium doped fibers will offer much for future studies of intravascular and intracardiac areas. For angiomas, especially the portwine type, the laser

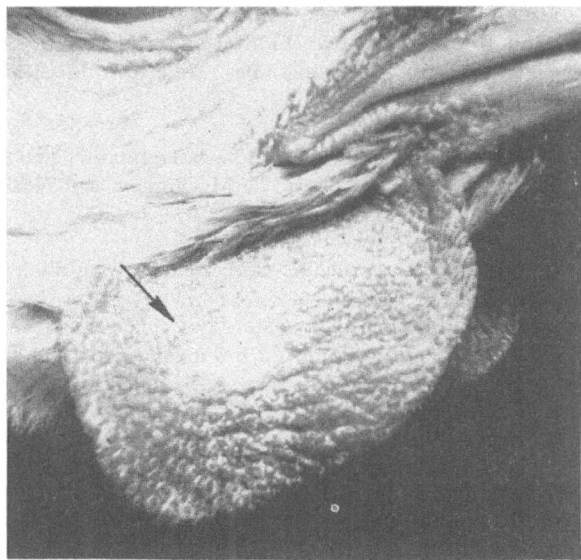

Fig. 13–5. Showing technique of Ritter for investigating the effect of the laser on angioma by the use of wattle of chicken. White area is impact of 49 joules/cm² from pulsed ruby laser.

appears to offer promise. For cancer of the blood vessels, high laser energy densities are needed. The role of the argon laser in these areas and in the treatment of angiomas is now under study with use of probes and fiber optics systems. Other CW lasers will be used in these studies.

REFERENCES

Baez, S. and J. P. Kochen: Laser induced microagglutination in an isolated vascular model system. *Ann. N.Y. Acad. Sci.,* **122**:738, 1965.

Cho, Sung Joon and D. Hershey: Blood flow in rigid tubes: Thickness and slip velocity of the plasma film at the wall. Dept. Chemical Engineering, University of Cincinnati, 1965.

Demis, J.: Personal communication.

Fine, S., E. Klein, and R. E. Scott: Laser irradiation of biological systems. IEEE Spectrum 81, April, 1964.

Forristal, T.: Personal communication.

Goldman, J., P. Hornby, and C. Long: Effect of the laser beam on the skin: Transmission of laser beams through fiber optics. *Journ. Invest. Derm.,* **42**:231, 1964.

Goldman, L., Robert T. Wilson, and K. W. Kitzmiller: Laser treatment of angiomas. *In press.*

——— and D. Richfield: Acquired angiomas. *Acta Dermato-Venerol.* **46**:177, 1966.

Goldman, L., Edmond J. Ritter, R. James Rockwell, Robt. Meyer, Bruce Henderson, and K. W. Kitzmiller: Exhibit-Laser surgery of angiomas with special reference to port wine angiomas. American Medical Association, June 18–22, 1967.

Goldman, S.: Personal communication.

Grant, Lester: Laser beam in tissue injury. The inflammatory process. Benjamin W. Zweifach, Lester Grant, Robert T. McCluskey. Academic Press, New York, 1965, p. 200.

Helmer: Personal communication.

Juhlers, Lennart: Vascular reactions in skin treated with fluocinolone. *Acta Dermato-Venerol.,* 44:322, 1964.

Klein, E.: Personal communication.

Kochen, J. A. and S. Baez: Laser induced microvascular thrombosis emboliza-tion and recannulization in the rat. *Ann. N.Y. Acad. Sci.,* 122:728, 1965.

Martens, T.: Personal communication.

Mertz, Edwin T.: Personal communication.

——— and John Y. S. Chau: Studies on plasminogen. *Canad. Jour. Biochem. & Physiol.,* 41:1811, 1963.

Naprstek, Z.: Personal communication.

Patterson, W. B. (ed.) : *Wound Healing and Tissue Repair.* Chicago, Univer-sity of Chicago Press, 1959.

Ritter, E. J.: The chicken comb and wattle as an experimental model for the therapy of hemangiomas. *Life Sciences,* 5:1903, 1966.

Rounds, D. E., J. Booker, F. F. Strasser, and R. S. Olson: The potentiation of gamma radiation with energy from a ruby laser. USAF School of Aero-space Medicine, Aerospace Medical Division. Brooks Air Force Base, Texas, Task #775702.

——— R. S. Olson and F. Johnson: The laser as a potential tool for cell re-search. *J. Cell Biol.,* 27:191, 1965.

Solomon, H., L. Goldman, B. Henderson, D. Richfield, and M. Franzon: Histo-pathology of the laser treatment of port wine lesions. *J. Invest. Dermat.,* in press.

Swarm, R. L.: Personal communication.

Yahr, W. Z., K. J. Strully, and E. S. Harwitt: Non-occlusive small arterial anas·tomosis with an neodymium laser. *Surg. Forum,* 15:224, 1964.

Yahr, W. Z. and K. J. Strully: Blood vessel anastomosis by laser and other biomedical applications. *J. Assoc. for Advancement Medical Instrumenta-tion,* 1:28, Sept./Oct., 1966.

Zweifach, B. W., L. Grant, and R. McCluskey: *The inflammatory process.* New York/London, Academic Press, 1965.

14

Laser Effects on the Skin

An important organ for investigative studies of the laser is the easily accessible skin. This organ provides opportunities not only for direct investigative studies on the effect of the laser on living tissue, but provides also living tissue for research on treatment. It is for these reasons that dermatologists have been interested in the laser.

SKIN STRUCTURES OF ANIMALS AND MAN

The skin has been a living model for research in all forms of radiation for many years. Studies of light transmission have revealed the complex optical properties of the human skin. Human skin is made of many different layers containing many different types of cells. The top layer, called the stratum corneum or keratohyaline layer, is a barrier layer which has amazing protective devices for the skin of man. Below this are the actively proliferating cells, the prickle cell layer; and then, the basal cell layer which contains the pigment producing the cells, or the melanocytes. Below this all is the rich network of connective tissue, blood vessels, sweat glands and nerves, and below this comes the fat tissues and the muscles. All these are very important for the absorption, transmission and reflectance of the incident beam of the laser. As Daniels indicates, these tissues are not homogeneous, contain absorbing and scattering pigments, particles and cell surfaces. He claims that some investigators forget that the skin is not just a photocell. In each and all of these complex tissues, light rays may be absorbed, transmitted and reflected at every level which they reach. Each different tissue has different optical qualities. The laser can be used as an investigative instrument for studies of the optical properties of the skin.

Since the skin of animals and man are different, they would of necessity have different light qualities, different optical systems affecting the incident beam of the laser on its surface. This is one of the reasons that it is difficult to carry over the analogy of the effects of the laser in animals on to man.

Experiments on animal skin do give but a rough idea as to the surface charring and penetrating effects of the laser beam as compared to man. The microscopic effects of the laser on the epidermis are different in man because of the thickness and especially because of the superficial quality of the epidermis. Perhaps this explains, for example, why neodymium laser may penetrate the skin of animals, whereas with the same energy density it will produce only a superficial effect on the skin of man and no deep penetration below the skin.

It is true that no laser should be used on man before detailed investigative studies are completed on animals. This is the procedure of any study of toxicity. If possible, even different species should be used. This is also the rule of studies of toxicity. Because of investigative studies in angioma, our laboratory is using miniature pig skin, although the vascularity of this is less than human skin. Moreover, Helwig had done extensive histological and histochemical studies of the effect of the laser on this animal. In addition, we have been using the comb and wattle of chickens for studies on angiomas. The decision as to when clinical investigative studies should start on man depends on the researcher's knowledge of what has been done in this and in related fields, what are the hazards, and, as usual, the calculated risks, and, consequently, what is needed for man. In laser research the urgency is there because of the great need for cancer treatment.

Hair is also of considerable importance, since it can absorb the laser and interfere with the passage of this beam through the other structures of the skin. In impacts of the hair, especially with the high energy laser systems, these occasionally burst into flame. Black hair has been used also by us as a test medium for studies of the value of proposed protective materials. The lens system of the laser is focussed on a black hair, and then various types of protective materials are put over the hair and the changes of the hair observed under moderate magnification with the stereo-binocular microscope and graded. Klein and his associates have studied reversible depigmentation of hair of mice to impacts of the ruby laser.

IMPORTANCE OF PIGMENT

Very early in the course of investigative studies on the skin of man, it was shown that pigment affected the absorption of the laser, and that the difference in response to the laser between Caucasian and Negro skin was significant. In investigative studies, the keratin of normal human skin has significant reflective and transmission qualities without showing any observable changes itself. This is not true, of course, for melanin. Re-

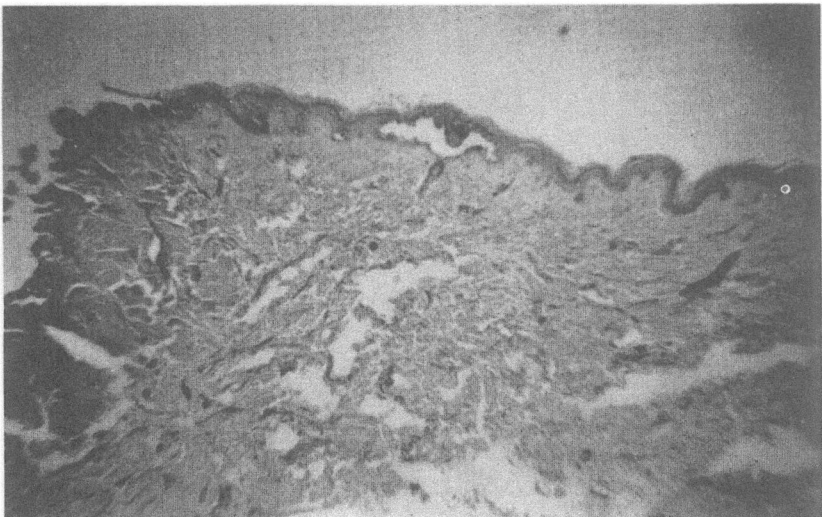

Fig. 14–1. Microscopic section of pigmented mole, mixed type, showing edge of crater following laser impact with skip areas in the epidermis where pigment foci (junctional elements) are localized. Hematoxylin-eosin, ×25.

flectance phenomena, that is, the study of reflectance on various types of skin surfaces after impact with the laser beam, has been studied. For these purposes, we have used the Lovibond Tintometer, a reflector developed in our laboratory to measure reflectance in the red-orange, 6000–7000 Å, with the peak at 6300 Å, and also the Reflecting Electro-spectrophotometer. Skin stripping with Scotch tape increased the intensity of the laser reaction. Everett and his associates found increase in ultraviolet erythema after removal of the stratum corneum. Studies have been made on the changes in dark film produced by the passage of laser beam through heavy bits of callous material. There were significant changes in the black film with very little changes, except for browning, in the callous material. Histochemical studies of keratin by the Barnett reaction showed only increasing color change in the target area of the laser. In Caucasian skin, with high energy laser impact, 50 joules or higher, it has been shown that very little change may be produced in the epidermis, but there may be thrombosis in the blood vessels in the mid-dermis. The skin, then, will show the effect of two natural pigments of the body in the absorption of the laser: melanin granules in the epidermis and hemoglobin of the red cells in superficial blood vessels. As regards carcinogenesis from laser radiation, the differences may be more important in human than animal skin, as has been shown in the studies of carcinogenesis by ultraviolet radiation in the skin of rabbits as compared to that of man.

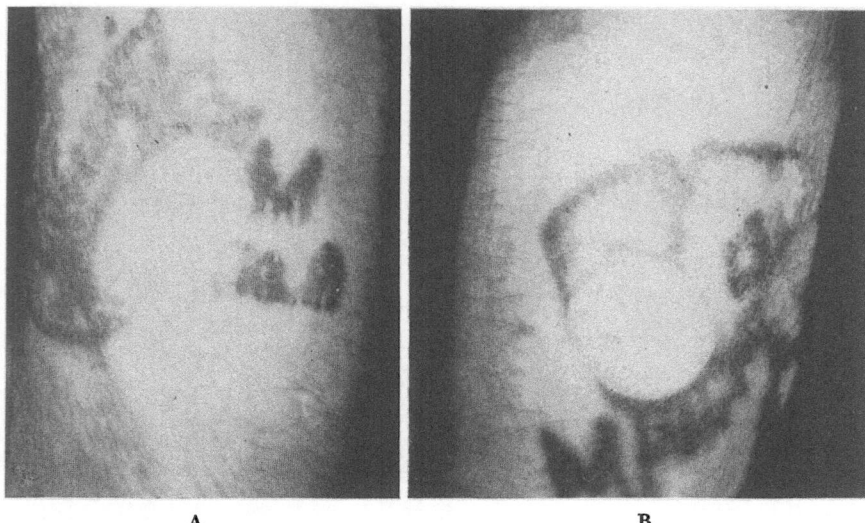

A   B

Fig. 14–2A. Depigmentation of tattoo, three years after impact with unfocussed beam, 200 joules/cm² energy density, from neodymium laser (*Eastman*). B. Depigmentation of tattoo, three years after impact with unfocussed beam, 152 joules/cm² energy density, from ruby laser (*Maser Optics, Inc.*).

INSTRUMENTATION

Laser radiation has been studied in the skin of man to date from low energy densities in various forms of laser into high energy densities, such as the 20,000 joules/cm² and Q-switching techniques of up to 100 megawatts in our laboratories. Lasers used have included ruby, neodymium, helium-neon gas lasers and high-power argon and carbon dioxide gas lasers. Laser beam may be used unfocussed or focussed by lens systems to penetrate in depth. We have focussed in cancer, in a woman, 3 cm. deep into the tissue. It may be used by transmittal through prisms, from reflecance from gold mirrors, and from passage, as indicated in Chapter 4, through quartz rods and segmented probes. With lower energy, transmission may be possible through fiber optics and plastic rods.

TECHNIQUES OF STUDY OF LASER IMPACTS

The changes observed in the skin following laser impact may be observed grossly in normal light or under Wood's light, ultraviolet light through a filter, by examination of the living skin by stereo-binocular and by capillaroscopy, and by detailed examination of the fixed tissue. In adjacent areas of the skin various energy densities of various lasers may be used, various focussing depths may be employed. All these tests can be done in adjacent areas so that the same microscopic section can reveal the different test sites.

The record of the impact may be made by photography. Photography should include, of course, pictures of the pre-impact as well as post-impact. The actual dynamic picture of the impact, with its plume, and the effectiveness of the plume traps used, may also be observed. Black-and-white, as well as colored film may be used. As indicated previously, in the chapter on laser photography, colored infrared film of Eastman has also been used to study the target zone and the extent of area reflectance from the target. Single-frame or high-speed photography may be used. The common manner in which the impact is studied is by the use of the open shutter of the camera in a dark room with the laser recording its own impact picture. Laser holography is being used to study the dynamics of the skin impact.

EFFECTS ON NORMAL SKIN

The acute effects of the laser vary according to the type of skin, the area of the laser impacts, energy and power densities and duration of the pulse, and the lens system. The excellent detailed pictures of the results on animal skin have been reported by Klein and his associates. The skin reactions of man considered here have been observed only for the past five years. Again, the old question arises of the necessity for the constant monitoring of each impact to account for the differences in the results of the laser impact. Pain is minimal after low energy impacts. With high impact of 200 to 300 joules exit energy, severe pain may be felt. Pain of the Q-switched impact, in nanosecond time, is still felt as "sticking" especially with the high peak power outputs.

Skin over hyperkeratotic areas, such as the palms and soles, shows relatively little effect. Dyes may enhance absorption of the laser beam. This has been reviewed in Chapter 9. However, if the dyes or India ink are just applied to the surface of the skin with little chance of penetration, the impact will remove the surface dye and produce little effect on the skin beneath.

The character of the surface skin reaction from the laser may be described as charring. This may be superficial or deep. Occasionally, there is bleeding especially when one is treating hemorrhagic areas. Later, this charred skin peels off leaving briefly a depigmented area which later pigments. In the Negro, temporary depigmentation may develop. As the charred area heals, it is very difficult to see it except for some discoloration of the skin when it is examined under Wood's light. If the coagulation necrosis induced by the laser is severe, scarring may develop. With our brief experience with high energy outputs, there appears

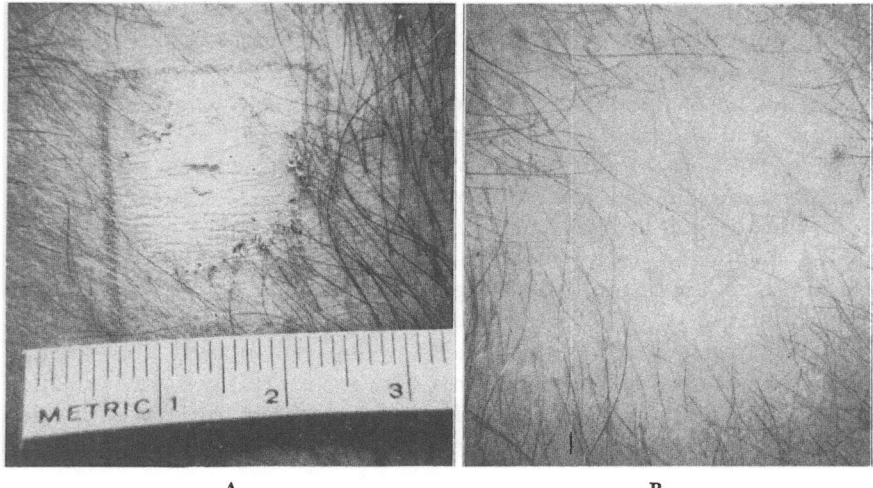

A   B

Fig. 14–3A. Impact on forearm of physician from unfocussed beam, 30 joules/cm²
energy density, pulsed ruby laser. **B.** Same area one and a half years later, showing only
slight atrophy.

to be initially excessive scarring, but eventually the scar flattens out. It
should be mentioned that there is no resemblance of the laser scar, during the past five years of research, to the scars following x-ray irradiation
of the skin. Few areas of telangiectasia have been seen in the laser treatment of superficial cancers of the skin and these are uncommon. Selective
subjects are being used as a study of the chronic effects of the laser.
Bits of their laser scars are excised from time to time to observe their
progress. Studies are also in progress of the effect of the laser on skin
collagen and on the possible synthesis of new macromolecules in the
dermis by the laser. The significant studies will be done with the ultraviolet laser. Collagen in the aging human skin is decreased and that
remaining is less soluble and more cross-linked.

As indicated in Chapter 13, studies of skin circulation after laser
impact have been done with impacts on the capillary bed around the
nails, on stripped, abraded skin; on ulcers, the ulcer base. Histologic
studies included alkaline phosphatase staining. As yet, no lymphangiog-
raphy studies have been done of skin circulation changes after laser
impacts.

Studies of chronic exposure have been done by deliberate exposure of
volunteers and by continued observation of laser personnel. Deliberate
chronic impact of the skin has been done only in the case of one investigator where it has been done more than 460 times with low energy
densities, 0.5 joules exit energy, in the period of the past four years in

which this laser investigator has worked in the laboratory. Initial lesions of this exposure were small pruritic papules which occasionally would leave a crust and at times a petechial spot. The immunologic studies of this reaction indicated for this individual that the skin became so-called hyperreactive. This means, it reacted to a laser energy density of a much lower magnitude, 18 joules/cm², than the energy density when the subject was originally exposed to the laser beam which was 100 joules/cm² of energy density to the pulsed ruby laser.

The following is a list of some factors which appear to affect the laser reaction in the skin:

1. Color—Caucasian or Negro color of the skin and vascularity
2. Hyperkeratosis or thickening of the skin
3. The characterist output of the laser used as regards type, exit energy, pulse characteristics, pulse duration and lens or transmission systems
4. Use of vital dye staining technique
5. The use of protectants, clothing, etc.

The enhancement of the laser radiation effect on the skin has been studied by the use of the various dyes mentioned previously, including various pigments in special solvents such as dimethyl sulfoxide (DMSO), dimethy acetamide (DMA) and dimethyl formamide (DMF). Copper iontophoresis has also been used to try to color the skin to get it to react more vigorously. Unless the pigment penetrates sufficiently into the skin, only superficial reactions will be evident.

Melanin still is the most effective absorber of the laser in the living skin, and the darker the skin is, the more intense the laser reaction will be. Smooth skin shows more absorption than thickened skin. The effect of the laser on melanin granules, on melanin biosynthesis and other factors of melanogenesis is now under study with histochemical, electron microscopic and electron spin resonance (ESR) techniques. Kobori and Toda have used the dopa reaction to study the effect of the laser on the melanocyte. The pigment cell was damaged, they believe, by heat absorption by the melanin granules of the dendritic processes.

An effort has been made to study in the living human skin the various multiphases of the laser reaction in tissue. Temperature production has been measured with thermistors by Wilson, in our laboratory, and by McGuff and Minton. In normal skin with an ambient temperature of 27° and unfocussed ruby laser 52 joules in the target area only 61° was reached and in tattoo skin 51.4°. ¼ inch away from the target areas, the thermistors measured, respectively, only 28.6° and 29.4° and deep in the skin in the target area 60.8° and 51.4° respectively. These thermistors, then, were inserted at various depths in tissue in and about the target area to record the depth changes in temperature through the passage of the incident beam of the laser. Temperature changes vary according to the position of the thermistor relative to the target. However, because of the slow response time of the thermistors, possible reflectance from the metal probes and absorption qualities in the deep tissues, it is a question whether these measurements truly reflect changes occurring in tissues. Our laboratory is developing special miniature, non-reflectant, rapid-response thermistors to attempt to overcome some of these difficulties. Vital dyes, fluorochromes (to evaluate plume traps) and tattooing, have also been used to study laser absorption in the living skin of man.

The use of barium titanate small pressure transducers have been used to study the transmission of the laser impact with ultrasonic wave formation through the skin and soft tissues as well as in cavities both in man and in animals.

The electromagnetic effects of the laser, so-called, have been studied by the use of x-ray film adjacent to the target area on the skin, various scintillation counters, ESR studies and detailed examinations of the pathology of the local reaction of the skin and the tissues below. The skin, then, offers facilities to answer the query that the only feature of the laser impact on living tissue is that of the thermal factor. Controls which have been used in these experiments have been the use of the electric needle providing the charring of the skin surface, the actual thermal cautery, the use of ultraviolet light, the use of xenon light itself, the plasma torch, and sun exposure.

These are but a few of the variables in the parameters of the laser reaction in skin. The laser, then, is another and very powerful instrument to study the mechanism of the effect of light on the skin in relationship to its inflammatory, melanogenic and carcinogenic properties.

Recently, attempts have been made in our laboratory to study the antigen specificity of burned skin in rabbits after the method of Kaplan and his associates, although they used no controls. If these experiments can be controlled suitably, they may offer some biological data for the differences of the laser and a purely thermal reaction. So far, we have been unable to show in immunized rabbits any specific laser burn antigen. Detailed studies on the morphologic, biochemical and immunologic differences between thermal and laser burns are continuing.

To date, observations of normal skin impacted by the laser as long as five years ago show only superficial scarring. There are no features clinical or histological, resembling that of x-ray burns. Longer periods of observation are needed.

Microscopic studies of the effect of the laser on tht skin are important. Microscopic sections offer a better technique to compare the differences in the skin of the local reactions in the same subject after different lasers and different modalities of radiation. The lasers used on normal skin were ruby, neodymium and argon gas lasers. On melanomas, carbon dioxide laser has been used. These various modalities of radiation used as controls included x-ray, Grenz ray, xenon light, electrosurgery with the high frequency electrical current, and the plasma torch. The closest resemblance, according to Richfield who studied the laser pathology extensively, is the resemblance of the laser destruction in tissue, called coagulation necrosis, to that of the electrical burn. Electrodesiccation and electrocoagulation should always be used for control studies, although it is difficult to select equivalent energy densities to compare the results. It is our impression, that although the surface changes may appear similar, the laser reactions are deeper, progressive and show more vascular damage. Telangiectasia, abnormal fibroblasts, homogenization of connective tissue and vascular occlusion pictures have not been observed, as yet, in the course of laser research. Histochemical studies have included PAS, Feulgen adenosine triphosphate, elastase, collagenase, and alkaline phosphatase, especially for blood vessel wall changes. An electron microscope is used also to define the minimal reactive dose (MRD) in the skin of man. The laser microprobe analysis of skin biopsies have been described previously.

Another technique which has been used to study the effect of the laser on the skin is the search for free radical formation by electron spin resonance spectrometry. This technique has been done by a number of different investigators on animal skin and living tissues before and after laser radiation. Our laboratory has recently done these studies on living human skin after laser radiation and other types of radiation. In the first series of experiments by Jacobi, only a pigmented birthmark showed significant signals. There was no pre-irradiation control. Although there have been scattered reports of free radical formation as detected on electron spin resonance spectrometry with the impaction of dark skin of animals, these studies, as yet, have not been corroborated in man. Perfection of the apparatus and increased facilities for clinical laser research will result in more extensive research to be done in that phase. The normal skin contains no free radicals, but pigmented tissues frequently show normal signals. Therefore, these tissues should have ESR studies before irradiation.

Table 1

RADIATION SPECTRUM FOR THE SKIN

2,900 A–3,200 A—Sunburn Spectrum
    MAXIMAL EFFECT 2,970 A
3,200 A–6,500 A—Usual Spectrum For Many Photodermatoses
    Induced By Drugs and Chemicals
3,000 A–6,500 A—Pigment Darkening Spectrum (Pathak, Riley, Fitzpatrick)

So, after the detailed animal investigative studies, the effects of various lasers can be studied on the living skin with due recognition of all the hazards there are involved in this particular study. Patients treated should be made aware of the possibility of these hazards, especially those of the unprotected eyes and burns to the skin. Electrical burns have occurred in laboratory personnel with unprotected high output investigative electrical equipment.

SKIN CANCER

Because of the accessibility and color, laser surgery can be used extensively in the field of skin cancer. The most significant treatments have been given for that black cancer of man, melanoma. Here, our laboratory has done laser operations even in delicate areas such as melanoma near the brain. It is too early to tell how permanent the effects will be, but enough data has been obtained to warrant continued investigative studies in patients with accessible cancer. Many other skin cancers besides melanoma have been treated. These will be described in Chapter 18. These include the pigmented and non-pigmented basal cell epitheliomas and squamous cancers. The other malignant tumors included also various types of malignant lymphoma as well as malignancies of the blood vessel. Efforts are being made now to stain transparent tumors to increase laser reaction with various pigments introduced into the tumor by direct application and injections. As we have indicated, where the target area has been adequate and the energy density has been adequate, these skin tumors have cleared and with good results. As with initial adequate therapy for any cancer, recurrences have developed. Basal cell epitheliomas have been followed for more than three years.

BENIGN SKIN DISORDERS

A variety of benign conditions have also been treated. These include warts of all areas (including the difficult periungual and plantar types), and shapes, polyps, tattoos and neurodermatitis or chronic thickening of the skin from nervous disturbances. All these have results indicating that controlled studies should be continued. The most striking results have been obtained with the removal of tattoos, especially with the Q switched laser. In tattoos there is opportunity for controlled studies of the effects of various lasers and various energy densities and depth focussing. Test areas are done first to determine efficacy and the intensity of scarring. Some patients with tattoos have been observed for more than three years.

Next in order of value for laser surgery are the port wine agiomas, for which, as indicated previously, there is no treatment. The laser beam produced significant superficial destruction of this angioma. Minimal to negative results were obtained in that chronic skin condition called psoriasis. Perhaps when the second harmonics are readily available,

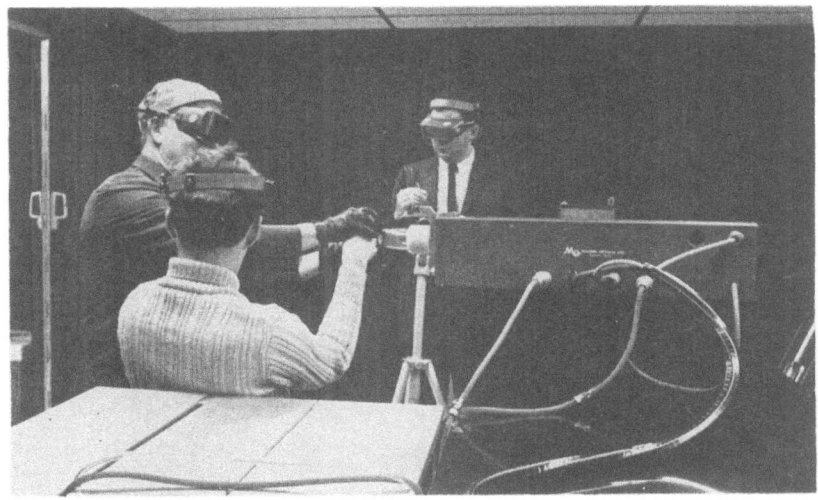

Fig. 14–4A. Treatment of tattoos. Showing arrangement of ruby laser for impacting tattoos on fingers of a patient.

Fig. 14–B. Showing use of experimental model argon laser on loan from Bell Telephone Laboratories.

Fig. 14–4C. Showing immediate effect of argon laser on "C," and "K" as untreated control.

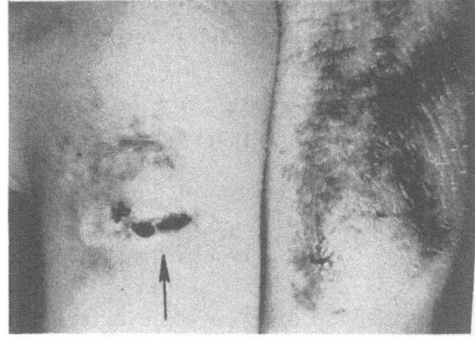

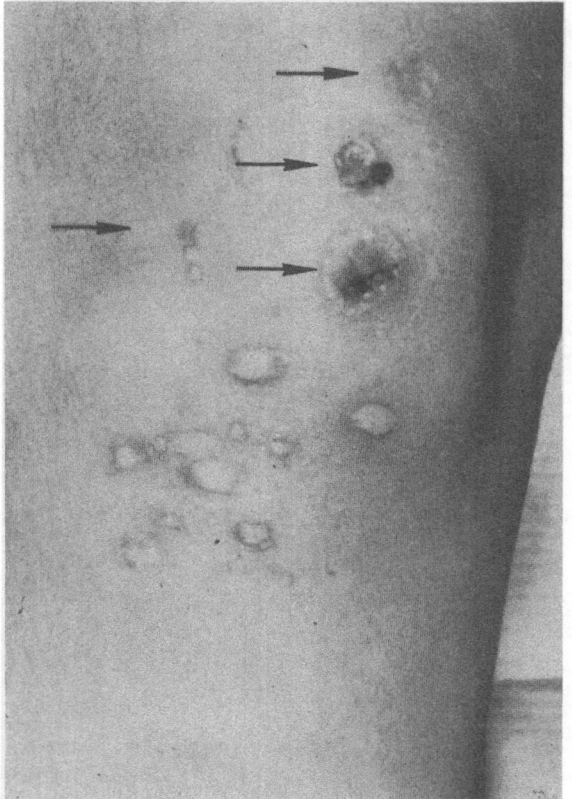

Fig. 14–5. Multiple warts on knee. After laser impacts, many flattened and some disappeared. Controls, untreated, were unchanged.

then these studies on psoriasis would be set up again. The question of the production of the ultraviolet light in skin through laser impact certainly should be under detailed investigative study. As indicated previously, the fluorescence of the skin from impact of the CW argon laser should have extensive investigation.

LASER TRANSILLUMINATION AS A BIOMEDICAL APPLICATION

The transmission of light through soft tissues has been used as a diagnostic technique in the field of medicine for many years. An intense beam of light is transmitted through skin or soft tissue in a dark room and such materials as foreign bodies, hard tumors, areas of infection, hemorrhage or even bone defects may show up as dark or light spots.

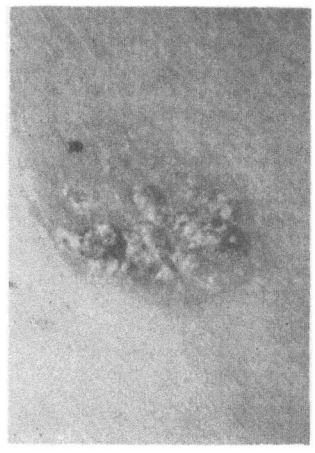

A

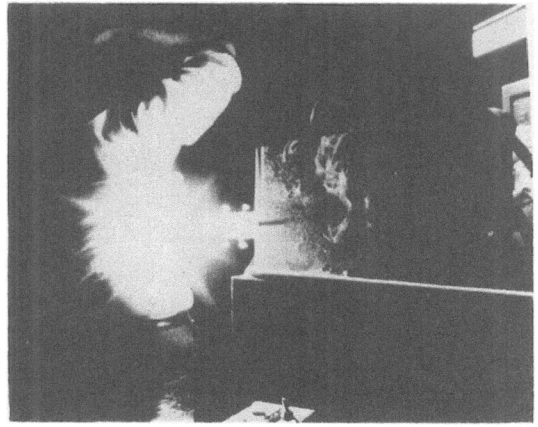

B

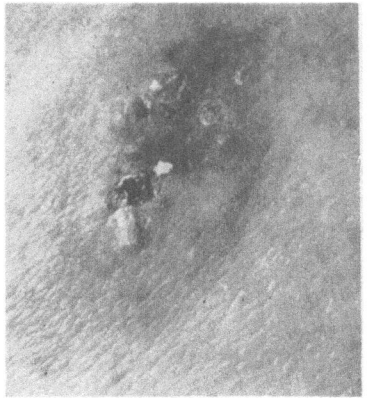

C

D

Fig. 14–6A. Seborrheic keratosis on chest of physician.

B. Showing impacts on chest using curved tapered quartz rod, 4mm in diameter, with 200 joules/cm² energy density from pulsed ruby laser. Black-and-white copy of colored infrared Ektachrome picture (*Eastman*) of impact.

C. Immediately after impacts.

D. Showing clear skin two years later.

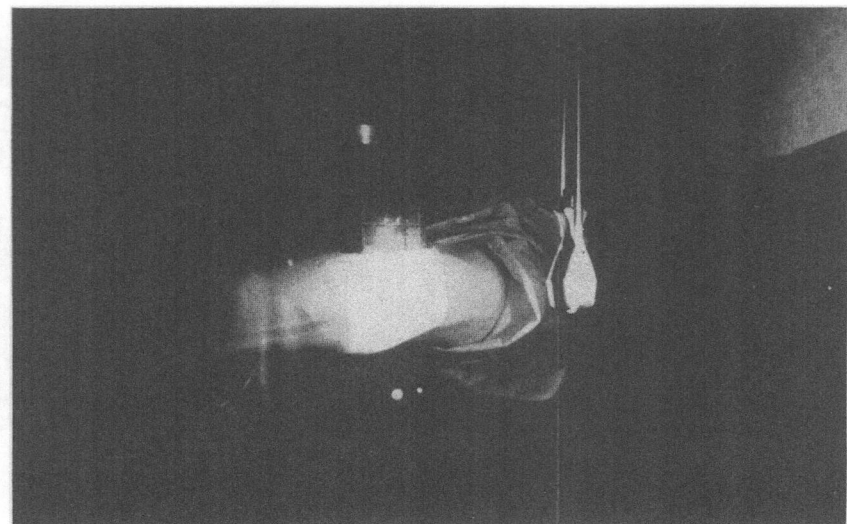

Fig. 14–7. Black-and-white copy of colored infrared Ektachrome (*Eastman*) picture of impact (85 megawatts peak power output, Q-switched ruby laser) on forearm for ESR studies. No free radical formation detected.

The technique is also used to detect infection in the sinuses of the head. Transillumination is used also in infants and children to transilluminate the head for abnormalities of circulation and abnormalities in the bony structure of the skull.

For transillumination, small flashlight holders are used with bulbs on the ends of extensions of the head. Also, light from large bulbs may be used and pictures may be taken with the use of flash photography.

The possibility of the use of the laser as a light source for transillumination has been under study by Friedman of our laboratory. The reasons for the use of the laser have been the greater intensity of the light source and the possibility of penetration of tissues similar to the use of so-called soft tissue x-ray photography technics. Recent developments in the field of x-ray mammography for the diagnosis of breast cancer with improvements in films and exposures have made this procedure widely accepted now.

In studies in our laboratory of the laser treatment about the head, neck and hands, efforts were made to try to evaluate the transmission of the laser beam through the tissues in these areas. Transillumination is of interest and significance not only because of the diagnostic possibilities, but because of the need to know the intensity of light which may reach vital structures such as the eyes. For example, in the treatment of cancers about the face with the laser, it is necessary to know how much light may

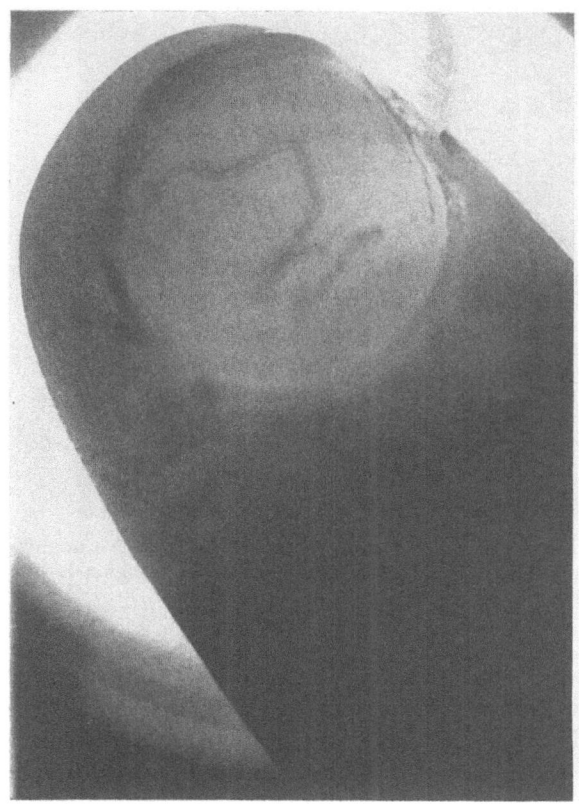

A

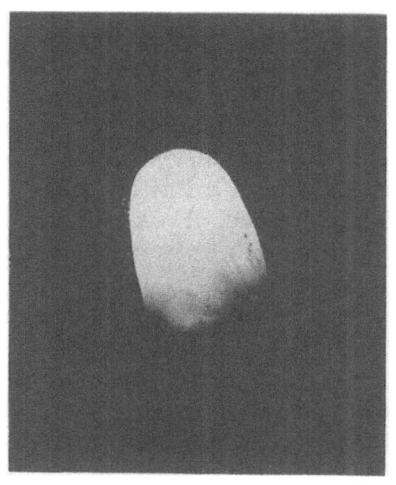

B

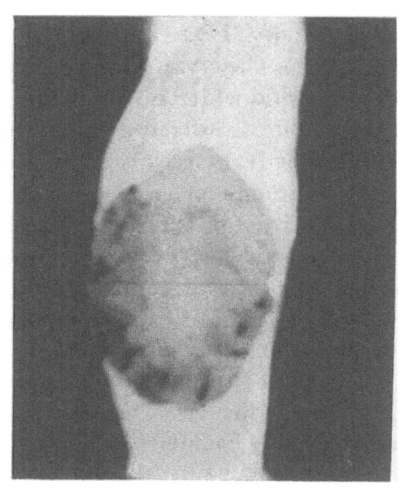

C

Fig. 14–8 (A) Transillumination of design on fingernail by pulsed ruby laser, 0.5 joules exit energy. Black-and-white copy of colored infrared Ektachrome (*Eastman*) picture (*Seth Friedman*). (B) Transillumination of finger with helium-neon 75 mw gas laser. (*Spectra Physics*). (C) Transillumination of excised mixed nevus with pulsed ruby laser 0.5 joules exit energy. Black-and-white copy of colored infrared Ektachrome (*Eastman*) picture (*Seth Friedman*).

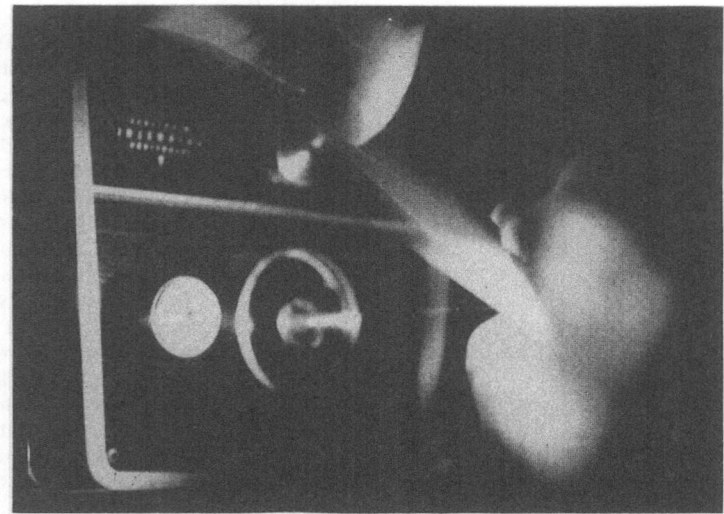

Fig. 14–8D. Transillumination of breast for suspected cancer with negative findings. Fifty-milliwatt helium-neon gas laser. Depth of transillumination approximately only 8–10 cms. Tongue blade used by gloved hand to hold the breast.

penetrate through the soft tissues into and around the eyes which are protected by heavy drapes and, in addition, by protective glasses. In laser treatment of the teeth, it is also of interest to know how much light is transmitted through the buccal cavity into the skull. In impacts of tumors of the chest wall, the laser beam may penetrate into the pleural cavity.

Continuous wave (helium-neon) gas lasers with only 1.0 milliwatt output were used to transilluminate fingers, nails and tattoo marks and excised tumors. These experiments were controlled with flashlight trans-illuminators. Observations were made in a dark room and photographed with black-and-white infrared film and colored infrared film (Eastman Kodak). The results indicated that this laser, with its weak output, offered no advantage over the transilluminators now available. High output helium-neon gas lasers, 50–85 mW (Spectra Physics, Inc.), are more effective for transillumination of soft tissues. Rockwell has suggested infra-red lasers for comparative studies with x-ray mammography for the diagnosis of cancer of the breast.

The next series of experiments were done by Friedman with a pulsed ruby laser, 0.3–0.4 joules of exit energy, with a lens system which focussed to a 1 mm. diameter spot, giving an energy density of 50 joules/cm². This laser beam was used on fingertips, nail deformities, excised tumors and tattoo marks. Since the pulse duration was on the order of milliseconds, the recording of the transillumination was done with black-and-white and with colored film in a dark room. Controls were done with flashlight transilluminators using similar films. In these studies, significant changes were observed beyond the regular flashlight illumination. Deformities of the nails were shown up in more detail. Dark masses were seen in excised moles. Additional experiments are continuing with use of the Q-switched laser and the neodymium laser and special photographic techniques.

PROTECTION

As indicated previously, for emphasis it is repeated that in the work in the Laser Laboratory, eyes must always be protected with glasses. Especially with high peak power outputs and high powered outputs of the ruby laser, unnecessary skin contact should be avoided by protection from the laser impact through the various kinds of shielding devices. The shielding margin about the lesion will be determined by the experience of the investigator. The protection may be secured by pieces of cardboard, by aluminum foil and by black cloth. In our hands, the white pieces of heavy but flexible cardboard, with their flexibility to be molded to different types and situations, have been the most suitable and practical laser protection of the skin. Reflectance from the skin surface may be significant and damage to veins may occur when these veins are superficial and the laser beam can impact them. Liquid protective creams and paints for the skin are still under investigation. Black gloves are often worn in holding young patients, such as infants with angioma, close to the laser beam. The effects of chronic exposure of more than four years is not known yet.

CONCLUSION

The laser, then, offers considerable opportunity for investigative studies in the effect of radiation on the skin. This opportunity should be utilized for more detailed and active work on living tissues. Studies on the skin of man should be continued, for his skin, optical system and reactions differ from those of animals. The laser as treatment for melanoma, other types of skin cancer and some forms of angioma and tattoos has shown effects which warrant continued controlled studies followed over a long period of time. Laser treatments should be done with due regard of the hazards.

REFERENCES

Daniels, Farrington, Jr.: Personal communication.
Everett, M. A., Waltermire, R. Olson, and R. Sayre: Modification of ultraviolet erythema by epidermal stripping. *Nature,* 205:810 (Feb. 20, 1965).

Fine, S., E. Klein, R. Scott, R. Seed, and A. Roy: Some effect of laser radiation on the skin of the Syrian hamster. *Life Sciences,* 1:30, 1963.

Friedman, Seth: Personal communication.

Giblin, T., K. Pickrell, W. Pitts, and D. Armstrong: Malignant degeneration in burn scars: Marjolin's ulcer. *Ann. Surg.,* 162:291, 1965.

Goldman, J., P. Hornby, and C. Long: Effect of the laser beam on the skin: Transmission of laser beams through fiber optics. *Journ. Invest. Derm.,* 42:231, 1964.

Goldman, L.: Dermatologic manifestations of laser radiation. *Fed. Proc.,* 24:1, 1965.

————: Derzeitige untersuchungen von dermatologischem interesse mit dem laser. *Der Hautaezt,* 1965.

————: Experiences with treatment of patients with high energy ruby and neodymium lasers. Presented at NEREM, November, 1965.

———— D. J. Blaney, D. J. Kindel, III, D. Richfield, E. Franke, and Ing: Pathology of the effect of the laser beam on the skin. *Nature,* 19:912, 1963.

———— ———— ———— E. K. Franke and Ing: Effect of the laser beam on the skin: Preliminary report. *The Journ. Invest. Derm.,* 40:121, 1963.

———— and K. W. Kitzmiller: Partial inhibition of the laser reaction in man by topical corticosteroids. *Life Sciences,* 5:2215, 1966.

———— and D. Richfield: The effect of repeated exposures to laser beams. *ACTA Dermato-Venerelogica,* 44:264, 1964.

———— R. Wilson, P. Hornby, and R. Meyer: Radiation from a Q-switched laser with a total power output of 10 megawatts on a tattoo of man. *Journ. Invest. Derm.,* 44:69, 1965.

————: Current status of the laser in Dermatology. *Dermatology Digest,* November, 1965, p. 47.

Helwig, E., W. Jones, J. Hayes, and E. Zeitler: Anatomic and histochemical changes in skin after laser irradiation. *Fed. Proc.,* 24:583, 1965.

Hoye, R. C. and J. P. Minton: The noble gas ion laser as a light knife. *Surg. Forum, vol. XVI* 5th Annual Clinical Congress 93, 1965.

Jacobi, T.: Personal communication.

Kantor, S. Z., I. Kaplan, E. Gruenberg, and S. Gitter: Antigenic properties of burnt skin. *Nature,* 207:540, 1965.

Klein, E., Y. C. Loar, L. C. Simpson, S. Fine, J. Edlow, and M. Litwin: Threshold studies and reversible depigmentation in rodent skin. Presented at NEREM, Boston, Mass., November, 1965.

————: Dermatological uses of the laser. *Gazette Medicale de France. In press.*

Kleine, Fines, Y. Laor, M. S. Litwin, J. Donoghue, and L. Simpson: Laser irradiation of the skin. *Journ. Invest. Derm.,* 43:505, December, 1964.

Kobori, T., and K. Toda: The effect of laser radiation on the epidermis. *Jap. J. Derm.,* 75:113, 1966.

McGuff, P.: Biomedical engineering aspects of laser radiation, NEREM Meeting, Boston, November, 1965.

Minton, J. P., C. D. Moody, J. R. Dearman, W. B. McKnight, and A. S. Kitcham: An evaluation of the physical response of malignant tumor implants to pulsed laser radiation, *Surgery,* 121:538, 1965.

Robinson, D. W., F. W. Masters, and W. J. Forrest: Electrical burns. *Surgery,* **57**:385, March, 1965.

Stratton, K., M. A. Pathak, and S. Fine: ESR studies of melanin containing tissues after laser irradiation. Presented at NEREM, Boston, Mass., November, 1965.

Tomberg, V.: Non-thermal effects of laser beams. *Nature,* **204**:868, 1964.

Wilson, R. G.: Personal communication.

15

Laser Effects on Internal Organs

Laser impacts may destroy tissue in any organ of the body. Because of the special interests, developments and importance, laser neurosurgery is considered in a separate chapter. In Chapter 15, the laser effects on organs other than the eye, the skin and the nervous system will be reviewed briefly.

The effects of the laser impact on internal organs are important for several reasons. First, it is desired to know how much effect there would be on the internal organs after use of focussed and unfocussed beams on various surfaces of the body. Do these impacts go into the organs below the surface? How deep do they go? As the laser beam penetrates through various tissues, are some affected more than others? Are some tissues not affected at all? Can there be accurate focussing deep into the organ from the surface? These are but few of the problems. It is obvious that in the development of protection programs against the laser, especially for the use of high energy and high power output equipment, the problem of intentional or accidental impacts of deep viscera is an important one.

There is another role of the use of the laser in regard to the impacts on viscera, namely, the use of laser surgery of these organs. For example, after incision of the abdomen or laparotomy, the abdominal organs may be exposed and then the laser used as a surgical tool in these areas. In these experiments, it is much better to study laser impacts in organs in situ than to use excised and dead tissues. The optical qualities of the organ are different under these different conditions. In the present stage of the development of the research program in this area, obviously the use of human material is limited to fresh material removed from the body. In our laboratory we have impacted with high energy ruby and neodymium lasers fresh material obtained from operations and autopsy of patients, including such tissues as the liver, spleen, stomach, colon, pancreas, aorta and bone. In all of these tissues, sharp lines of demarca-

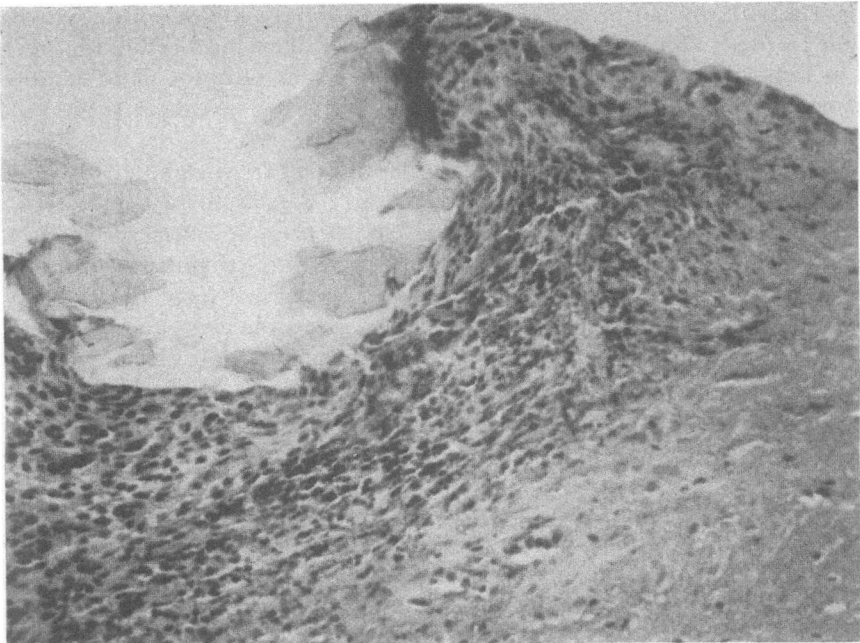

Fig. 15–1. Microscopic section of pulsed ruby laser impact of fresh autopsy specimen of man; 60 joules exit energy showing deeply punched out crater of bowel with sharply demarcated necrotic base hematoxylin-eosin ×90.

tion of the laser-induced necrosis were evident especially on microscopic examination with routine stains and with special histochemical techniques.

LASER INSTRUMENTATION

All types of lasers have been used in the study of effects on internal viscera. These include low and high energy ruby, neodymium, normal and Q-switched modes. Of special interest and concern is the effect of the high energy and high output ruby and neodymium lasers. Tht recent developments of the high output argon laser has shown this to be an effective surgical tool especially for those viscera which bleed considerably, such as the liver, the spleen and the lung. This one type of laser will continue to be used extensively in the field of laser surgery. The major problems for the argon laser are, of course, the present lack of flexibility of the instrument and the complex program of eye protection for the operating surgeon. Special fiber optic systems and special flexible probes will make the argon laser more flexible and special glasses and screens will protect the surgeon. Some laser probes and fiber optic systems will be used with endoscopic instrumentation. The carbon dioxide laser can reduce bleeding on impact with high power outputs. Considerably more development will be necessary before it can be used for surgery in man.

Because of the interest in reaching depth areas in viscera, lens systems will have to be used with the lasers to carry the beam into the viscera. These problems in focussing in depth have been reviewed previously. In this field, also, the use of straight and curved quartz rods for transmission of the laser beam will be employed. These will have the value of making the laser more flexible to reach the accessible areas and to deliver the beam exactly on the target area. Special rods, as indicated previously, have high outputs and transmit the beam only at the exit end of the rod and, therefore, without danger to tissues about the rod. These rods, then, may be inserted deep into viscera, also, to develop special techniques of impact.

It is hoped that the development of the junction-diode type of laser with the higher output and with miniaturization of equipment will make this possible to be used in body cavities for selective local action. This is not possible at the present time.

ANIMAL EXPERIMENTS WITH PERFORATION
OF ABDOMINAL VISCERA

Early in the experiments of Fine and Klein, it was noted with impacts of the abdominal surface of mice with pulsed ruby laser exit energy of 75 joules that it was possible to get lesions of the underlying viscera such as the liver and the intestine. It is also important to note that tissues intervening between the skin surface and the viscera, such as the layers of the peritoneum, were not damaged by the laser. This is due to the differential absorption of the laser beam in pigmented areas. Increasing energy levels to 300 joules exit energy causes much more severe changes, including blood in the peritoneal cavity, lesions in the kidney, spleen and pancreas. Similar lesions were observed with the neodymium laser with energy levels exceeding 500 joules.

With the use of high peak power radiation Q-switching techniques, damage to viscera deep into the peritoneal layer was present in the liver, intestine and stomach. Again, the alteration and sequence of damage in normal tissue was noted. This differential absorption of the laser beam as it passes through various tissues certainly is an argument against pure thermal quality of the laser energy.

Experiments in our laboratory by Gentele, McGuff and Henderson have shown similar results with involvement of abdominal viscera in impacts of the shaven abdominal skin of animals. The neodymium laser with suitable equivalent energy densities had deeper penetration qualities than the ruby laser. For the penetration of the abdominal viscera of dogs, much higher energy densities were required.

In the treatment of pigmented lesions of the skin and soft tissue of man on the chest and abdomen, exit energies of ruby laser to 300 joules and neodymium laser to 200 joules have shown no evidence of intra-abdominal or intra-thoracic penetration.

THE LIVER

Because of the special interest in laser surgery of the liver, this organ will be considered in detail. Our initial experiments in impacts of the liver were done with blocks of fresh human liver obtained from autopsy. In these blocks, the effects of topical injection of various dyes were studied, the use of depth-focussed doses and the insertion deep into the liver of special quartz rods. The effect of doses focussed 3 cm. in depth have been done in these blocks of liver, and the coagulation necrosis induced has been studied microscopically. It is possible with the use of closely spaced laser impacts to get in the same section of tissue unfocussed, surface-focussed and depth-focussed lesions, all for comparative microscopic analysis. As indicated previously, studies of isolated bits of dead tissue do not have the same value as those of the liver in situ.

The most extensive experiments on the effect on the liver have been done by Minton, Ketcham and Mullings. Perhaps the most interesting portion of the studies has been done with the use of high energy lasers on the primate liver. These experiments were preliminary to the use of the laser in destruction of liver cancer or hepatomas involving the liver. With detailed studies of serial liver biopsies after high energy impacts on the liver, Mullings and Hinton found that there was uncomplicated recovery of these lesions and survival of the primate although microscopic foci of cancer persisted in the deep scars. This is of great interest because of extensive necrosis which may be induced by unfocussed and focussed impacts on this highly vascular organ. Fifteen-hundred joules exit energy was maximum energy used in their experiments with the high energy ruby and neodymium lasers. In our laboratory, McGuff and Henderson have studied acute and chronic lesion after laser impact in animals and have also found healing of such tissues with a significant amount of scarring and survival of animals.

The recent uses of the argon laser as a surgical instrument for the liver by Hoye and Minton and by Brown and Henderson in our laboratory, have shown that this is an effective surgical tool in this organ which is so vascular and consequently bleeds so easily. Additional experiments are continuing here with regard to the use of plastic adhesives to suture areas after removal of portions of tissue and gold leaf to prevent second-

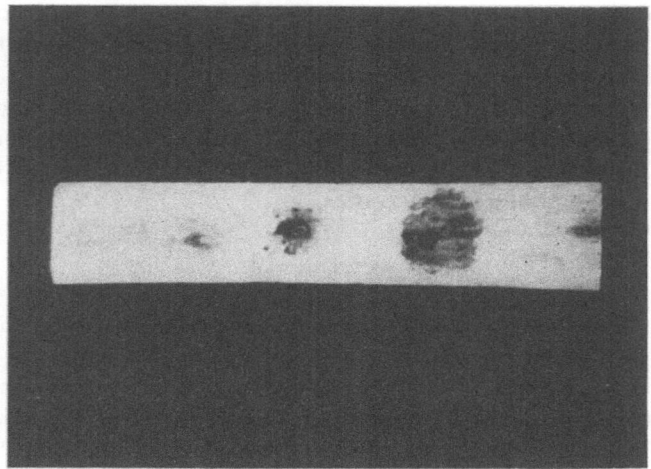

Fig. 15–2. Showing effect of energy densities of 100 joules/cm², 200 joules/cm², 300 joules/cm² from pulsed ruby laser on section of polished cortical bone. (*Collabration research with Dr. Robert O. Becker, Chief, Orthopedic Section, Veterans Administration Hospital, Syracuse, New York.*)

ary hemorrhage. All these preliminary animal experiments indicate that there is a great future with the use of this laser in liver surgery, providing high power outputs can be used.

BONE

Our experiments with bone show considerable charring of fresh ribs when impacted in situ or shortly after removal at autopsy. The charring extended from the surface down into the cavity of the bone. Repetitive impacts of the laser could produce a hole in the bone, but high energy outputs were required. Experiments have been started with Becker and Bassett on the laser impact of polished bone sections and apatite on bones in situ to determine the effect of laser radiation as regards histologic section and EPR studies. Again, high energy densities were required to drill the bone. This has been possible with carbon dioxide lasers to high power output 10–25 watts. Neodymium lasers were used also to try to study infrared absorption of bone. Energy densities used on dog femur have included 17,000 joules/cm² surface-focussed, normal mode pulsed ruby laser; 8,500 joules/cm², focussed 0.5 cm below the surface; and similar energy densities studies with the neodymium laser. With the easy accessibility of bone and its storage in dimethyl sulfoxide (DMSO) and with the fact that the bone can function of itself as a junction diode laser, a new field of laser research has been opened. Studies of the effect

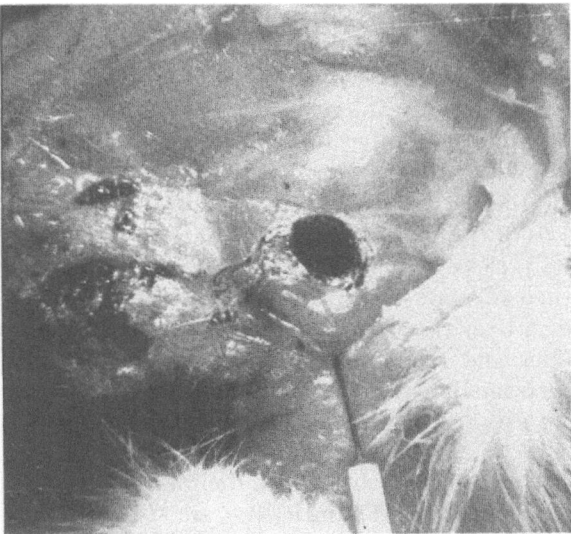

Fig. 15–3. Carbon dioxide laser impact on muscle tissue of guinea pig showing thermocouple to record temperature rise in tissue.

of laser radiation on the bone should be compared to that of ultraviolet, infrared, xenon and ionizing radiations.

The laser has been used as a probe for microanalysis of undecalcified bone tissue by Lithwick, Cohen and Healy. The vaporized material was analyzed spectrographically for calcium, phosphorus, fluorine, nitrogen, zinc and magnesium. Specific areas of the bone can be selected by the laser microprobe and the analysis restricted to this particular area. The whole subject of spectrochemical analysis of pathologic bone by the laser microprobe is an interesting one for future investigations. Studies of the possible production of fat emboli from laser impacts in the bone should be done.

MISCELLANEOUS ORGANS

Other studies done in relationship to the laser have been with the bladder, by Parsons in dogs, and Mulvaney in our laboratory with regard to the possibility of the use of laser in urology. The development of laser endoscopic instruments is of interest, but for operative surgery transmission rods will be required which can transmit relatively high energy laser beams. In our laboratory, experiments are being done with the use of long, special quartz rods for endoscopic laser surgical techniques.

As detailed previously, laser fiber optics endoscopy has been limited because of the inability of the fiber optics now currently available to transmit adequate outputs of the laser. The continued developments of special quartz rods may relieve some of these difficulties, but still would not make for flexible instrumentation. At present, the flexible fiber optics can be used as light pipes for direct visualization of deep lesions. Schechter, in our laboratory, is constructing a colpomicroscope for direct visualization of the cervix so that it may be possible to direct laser transmission rods or special flexible probes to the superficial precancerous lesions, especially those which have been stained selectively. This dye staining will then increase absorption of the laser beam.

With regard to laser research on calculi, this concerns not only calculi from the bladder and kidney, but also from the biliary system. Laser microprobe spectroanalysis can be done to assist in the identification of these stones. Laser surgery may be used to fragment or vaporize dyed stones.

Stahle and Högberg have impacted the inner ear of pigeons with Q-switched ruby laser. With Shumrick we are studying the applications of the laser, especially with quartz rods and fiber optic systems in the field of maxillo-facial surgery. This includes drilling of the mastoid bone, operations in the middle ear and larynx and treatment in patients of lesions of the lips, gums and tongue.

As mentioned previously, electrocardiographic studies have been done on impacts of the laser over the precordium in dogs and ECG studies are being done in man following the impact of superficial lesions of the skin and soft tissue over the precordial area. So far, no abnormal reactions have been found with unfocussed and surface-focussed pulsed ruby and neodymium lasers.

LASER TRANSILLUMINATION OF
INTERNAL ORGANS

The technique of laser transillumination of soft tissues of the skin has been presented previously. Attempts were made to determine whether with instrumentation now available it would be possible to do laser transillumination of viscera, such as the breast and the chest and the sinuses of the skull. Ruby lasers with outputs of 70 joules were used, neodymium lasers with outputs of 200 joules. The recording film was placed on the exit areas of the opposite surface. For example, for attempts at laser transillumination of the chest, the unexposed film in the darkened laser treatment room was put at the back of the chest if the impact was at the front, and on the front of the chest if the impact was at the back. The edge of the film was protected from any reflected radiation in this darkened room. For the use of bone, films were placed at the other side of the extremity; for use with the breast, underneath the breast; for the use of the front or top of the skull, in the posterior aspect. With various colored films and with infrared color and black film it was not possible in these areas to detect any transmitted laser radiation. In some of the experiments, photodiode tubes were used also, and also detection of light was found. Experiments with small barium titanate transducers are still under investigation. Q-switched lasers have not been used, as yet, in these experiments.

CONCLUSIONS

The results of penetration of the overlying skin into deep internal viscera are quite different in animals than in man. This is done easier in animal tissues. With the increasing development of high energy and high output lasers these internal organs must be protected also. The recent development of high output cw argon and carbon dioxide lasers to function as light knives for operation on internal viscera is of great interest. This opens up a new field for bloodless surgery in important organs. What is needed, as usual, are long-term studies to follow the damage and the healing in these viscera. Continued research will have to be done to determine whether laser surgery will be effective not only in removal of portions of these organs, but also the use of transplantation of organs. Very little has been done in laser surgery of the internal organs, and with the few preliminary studies done all fields offer great opportunities, as for example, liver, lung, buccal cavity, bone and the genitourinary system. The laser microprobe also offers a field of investigation for the spectrochemical analysis of the normal tissue of these internal organs, the tumors which it develops and, also, for the various calculi that may occur in these organs.

REFERENCES

Barnes, F., J. Daniel, and K. Takahashi: Laser surgery and biological damage. Presented at NEREM, Boston, Mass., November, 1965.

Bassett, C. A. L.: Personal communication.

Becker, R. O.: Personal communication.

Brown, T. E.: Personal communication.

Fine, S., T. Maiman, E. Klein, and R. E. Scott: Biological effects of high peak power radiation. *Life Sci.,* 3:209, 1964.

———— E. Klein, W. Nowak, R. E. Scott, Y. Laor, L. Simpson, J. Crissey, J. Donoghue, and V. E. Derr: Interaction of laser radiation with biologic systems. I. Studies on interaction with tissues. *Fed. Proc.,* 14:35, January–February, 1965.

Friedman, S.: Personal communication.

Goldman, L.: Comparison of the biomedical effects of the exposure of human tissues to low and high energy lasers. *Ann. N.Y. Acad. Sci.,* 122:802, 1965.

Goldman, L., Donald Shumrick, J. R. Rockwell, and Robert Meyer: Investigative studies with the laser in maxillo-facial surgery. Presented at the Section of Maxillo-facial Surgery of the American Medical Association, June 21, 1967.

Hoye, R. C. and J. P. Minton: The noble gas laser as a light knife. *Surg. Forum,* 16:931, 1965.

Lithwick, N., J. Cohen, and M. Healy: Microanalysis of bone by laser probe. First Annual Biomedical Laser Conference, Boston, June 14, 1965.

———— M. Healy and J. Cohen: Microanalysis of bone by laser microprobe. *Surg Forum,* 15:439, 1964.

McGuff, P. and B. Henderson: Personal communication.

Mullings, F. and J. P. Minton: The effect of multiple high energy laser pulses on the primate liver. First Annual Biomedical Laser Conference, Boston, June 17, 1965.

Schechter, E.: Personal communication.

Shumrick, D.: Personal communication.

Stahle, L., and L. Högberg: Laser and the labyrinth. *Acta Oto-Laryngologica,* 60:367, 1965.

16

Laser Neurosurgery

Recently, workers in neurosurgery research have studied the effect of laser radiation in neural tissue. Interest in these studies evolved because the laser provided a new form of energy characterized by a precisely controlled beam with power density which could be accurately varied. Furthermore, the recent development of high repetitive rate, water-cooled laser systems has been of unmistakable significance to this advance. As will be shown later in the chapter, low energy densities may be used to cause the desired and precise destruction in areas of the brain, spinal cord, and peripheral nerves which permits the use of relatively simple laser systems.

Investigations by Brown and his coworkers from this laboratory have shown that the pulsed ruby laser can be applied safely in highly critical areas of nervous tissue. Investigations are being carried out by Brown and his coworkers, Stellar, Fox and Earle, and by Rosomoff to study the immediate and long-term effects of laser radiation to a variety of neural structures. The application of laser energy to experimental and clinical neurosurgery is in its infancy, but will undoubtedly accelerate.

LASER INSTRUMENTATION

The pulsed ruby laser is most frequently used for laser neuro-surgery research at this time. Liquid nitrogen-cooled and water-cooled systems have been utilized for these studies. Beam delivery has been accomplished by three methods: (1) unfocussed beam, (2) tapered quartz rods and (3) simple optical lens systems. Flexibility of mounts for the laser system remain quite limited, although a mobile unit has been utilized in our laboratory that gives some maneuverability. More flexible devices are required for accurate positioning in order to approach any intracranial structure. Preliminary investigations in this laboratory and others have been initiated to study the effects of laser energy in neural tissue from the Q-switched ruby laser, the neodymium laser, and the argon laser in neurosurgery research.

EXPERIMENTAL STUDIES IN LASER NEUROSURGERY

When the effects of laser irradiation on intracranial neural structures is considered, one must emphasize whether the impact is delivered to the intact head of the experimental animal or directly onto the exposed neural tissue. As will be shown, the effects differ markedly and it appears likely that the mechanisms are different as well. Delivery of a physical force into a confined cavity (such as laser bombardment of the intact skull in an experimental animal) could unleash a tremendous force at the site of tissue vaporization which would then be disseminated throughout all structures contained within that cavity. However, when such a cavity is opened and laser impacts are directed onto the enclosed structures, the dissemination of such a force would now appear quite unlikely since the confines of the cavity have been destroyed. In essence, the former situation could result in uncontrollable destruction for cranial contents. This has been well documented by several investigators. The latter situation (direct impacts) would theoretically permit precise control of the destructive forces of the laser beam.

LASER IMPACTS OF THE INTACT SKULL

Fine and Klein originally described the effects of laser radiation on neural tissue, following delivery of impacts of a pulsed laser beam to the frontal area of the intact skull in mice. These investigators utilized exit energies of 100 joules with a pulse duration of 1 millisecond. Many of the subjects expired shortly after impact or within 24 hours. The underlying brain showed extensive damage with petechial hemorrhages noted throughout the brain tissue as well as in the brain stem. Beneath the impact site a cone shaped area of total cell destruction was usually found. Similar experiments have been performed by Earle and his associates at the Armed Forces Institute of Pathology and by Brown and his co-workers.

These results underscore the drawbacks to use of laser energy in such a manner. Damage is severly uncontrolled and lethal, if sufficient energy is used. However, such findings have a useful application. Previously, it had not been possible to develop a model for uniform production of closed craniocerebral trauma. This experimental technique is vital for neurosurgery research to permit controlled investigation of the etiology, pathogenesis and treatment of head injury. Furthermore, very little is known of the mechanism of cerebral edema and, as will be shown, the pulsed ruby laser can be utilized to provide such an experimental model. Brown and his coworkers have utilized the pulsed ruby laser beam to produce uniformly subdural and subarachnoid hematomas and cerebral edema in the laboratory rat. Exit energies were varied from 25 to 400 joules and delivered as single unfocussed and focussed laser impacts to the intact scalp as well as to the exposed skull of such animals. Results demonstrate that impacts to the intact scalp at the lower energy levels (below 100 joules) induced subdural hematomas and cerebral edema in approximately 50% of the animals. It is interesting that skin pigmentation in these animals alters the incidency of intracranial injury. When the head of a white laboratory rat is impacted with 100 joules or more (focussed beam) cerebral edema and/or subdural hematoma results. However, when the same focussed impacts were delivered to the heads of black hooded rats the incidence of subdural hematoma and serious intracranial injury fell to less than 50% at comparable energy densities. The naturally occurring pigment of the black skin of the hooded rat probably permits maximal absorption of the laser beam at the skin surface with its dissipation in this rather loose tissue layer of the scalp. Consequently, the incidence of significant intracranial damage is greatly impaired when the path of the laser beam is interrupted by pigmented substances.

When subdural hematoma is produced, neurological deficits appear, including generalized convulsions, coma, progressive hemiparesis and finally, hemiplegia on the side of the body opposite the site of the subdural hematoma. Cerebral edema ensued within four hours after the occurrence of the subdural hematoma and it progresses over the next 72 to 96 hours. It is important to note that the size of the subdural hematoma did not change for at least four to seven days, whereas the degree of cerebral edema developed to a marked level within 70 to 96 hours. During the time of onset of cerebral edema, mortality rate of the experimental subjects increased and progressive deterioration was noted in the subjects' clinical condition, strikingly similar to the progressive downhill course of similar clinical cases of closed craniocerebral trauma. Microscopically, the hematoma was found to reside in the subdural space on occasion, and more commonly in the subarachnoid space with a wedge or saucer-shaped area of total cortical necrosis immediately beneath the clot. On occasion, subarachnoid hemorrhage was found around the base of the brain and very rarely in the intraventricular compartment. When 400 joules was delivered to the intact scalp of the laboratory rat, immediately respiratory paralysis ensued and cardiac arrest followed within 10 minutes. The brain showed diffuse cortical necrosis over the dorsum of the hemispheres and a rim of early subdural and subarachnoid hematoma. Presumably, the rapid death of the animal did not permit development of the hematoma. By utilization of appropriate energy levels, the entire spectrum of closed craniocerebral trauma can be produced on a scale satisfactory enough to provide a useful model to study the course of events following head injury. More significantly, such a technique offers promise as a screening method to evaluate therapeutic agents rapidly for treatment of cerebral edema resulting from physical injury to the brain. This latter point must be emphasized, because the onset of cerebral edema appears to be a factor of equal importance demanding therapy as removal of the clot itself for improved morbidity and survival rate in such patients.

In summary, the use of pulsed laser systems to reproduce closed craniocerebral trauma offers an excellent experimental tool to study more closely the syndrome of head injury in man. Therefore, the applications of laser appear equally fruitful in the area of basic research as well as clinical application.

LASER IMPACTS AFTER CRANIECTOMY

In contrast to the above described widespread intracranial damage, discrete lesions can be induced by laser impacts when the intervening skull plate is removed and the cerebral tissue impacted directly. Such lesions are precise and easily controllable if appropriate energy densities are employed. Brown and his co-workers have investigated the effects of such impacts upon cerebral cortex for more than a year, and Earle and his associate and Stellar have performed similar investigations. An intensive investigation was carried out by Brown and his co-workers to study the effects of laser impact after unilateral or bilateral craniectomy in mongrel dogs and monkeys (Machacus Iris). The laser beam was delivered to the target site as a focussed beam (in which the effects of a divergent, surface-focussed and convergent beams have been studied) and through appropriate light guides such as the tapered quartz rod. Over 300 acute impacts have been studied on cerebral cortex and certain findings are worthy of comment. The lesion caused by focussed impacts was discrete, consisting of rounded lesions varying from one to 2.5 mm in diameter, depending upon the energy level utilized as well as beam convergence employed. Five joules delivered from the pulsed ruby laser caused punctate blanching with small areas of subpial hemorrhage. When energy levels were varied between 10 and 40 joules (focussed), the lesion became well-defined and uniform in appearance, characterized by a central core of complete necrosis and a marginal rim of variable hemorrhage. Brain sectioning revealed that the lesions were confined within 2 mm. of the surface. Rarely, such necrosis extended through the entire cortical layer. Evidence of distant brain damage from focussed impacts was not observed in the experiments. Microscopically, the lesion appeared similar to a hemorrhagic infarct and was cone-shaped, with its apex pointed toward the subcortical white matter. Vessels within the lesion were charred and dehisced. Intravascular thrombosis was seen commonly. Presumably, the thrombosis accounted for the limited hemorrhage. For approximately 1 mm. beyond the edge of the lesion vessel destruction was seen which was similar to that within the central core, although the intervening brain tissue showed little or no cell death and little edema. It appeared that vascularity is an important determining factor for the interacting of laser radiation in a lightly colored tissue such as brain tissue. To a certain extent, pigment granules found locally are obviously important to the laser effects. The neuropathologic studies carried out by True, permit the tabulation of characteristics in a laser lesion of cerebral cortex.

CHARACTERISTICS OF LASER LESION IN NEURAL TISSUE

Blood vessel effects—Pronounced destruction with dehiscence and intravascular thrombosis. Laser interaction with neural tissue enhanced in area immediately surrounding local vascular bed.

Difference in cell susceptibility (neuronal tissue) —Of the cell components, neurons appear very susceptible to laser irradiation. Furthermore, the giant pyramidal cells of Betz are more susceptible than the smaller association neurons of cerebral cortex. Astrocytes appear to be relatively resistant to laser irradiation. However, chronic studies reveal a unique and marked delay in astrocytic healing of the laser lesion.

Lesions are precise and well controlled—Distant brain damage was not found in more than 300 impacts with focused ruby laser beam when the intracranial tissue was impacted directly by the laser beam.

Susceptibility of gray matter is greater than white matter susceptibility— Highly cellular tissue (gray matter) proves very susceptible to necrosis from laser bombardment, whereas fibrillary tissues (such as subcortical white matter, the internal capsule, the spinal cord or nerve trunks) are relatively resistant to laser interaction. A significant difference exists in threshold of each type of tissue to laser energy density to permit some selectivity of localization of the laser effects in a mixture of such tissues.

In long-term studies for three months after laser irradiation, there have been no detrimental side effects observed directly attributable to laser irradiation. The preciseness of the lesion and its accurate control in terms of energy density offer a promise of future clinical application of lasers in the field of neurosurgery. Since laser irradiation affects gray matter more easily than white matter, it appears feasible to utilize this principle in critical areas of the brain, such as the basal ganglia, and yet spare the adjacent internal capsule.

Forty-eight thalamotomies have been performed in 16 mongrel dogs

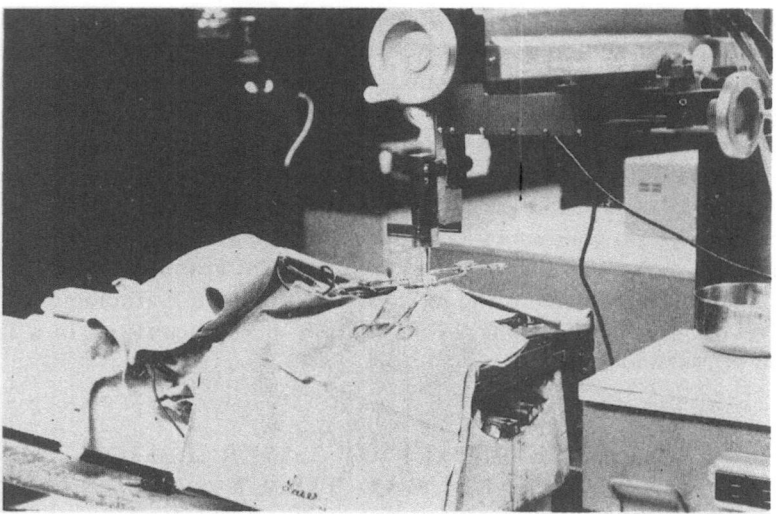

Fig. 16–1. Technique of Brown for the operation of laser thalamotomy in the dog. The tapered, curved quartz rod attached to the head of ruby laser is used to transmit the laser beam to the exact target in the brain.

and in 8 Machacus Iris monkeys by the following technique. A straight, tapered quartz rod 142 mm long with final exit diameter of 2 mm and of 4 mm was utilized in this series of experiments. The pulsed ruby laser beam was delivered at energy levels of 8 joules, 15 joules and 30 joules through the quartz rod into the depth of the brain substance. Average pulse length was 1.7–2.2 msec. No significant escape of laser radiation appeared to occur along the course of the rod except at its end tip. At the exit face of the rod, a discrete, well-rounded lesion, measuring from 2 to 8 mm (from the lowest and highest energy levels employed, respectively) resulted. Even where lesions were placed immediately adjacent to the internal capsule there was no clinical evidence of damage to the adjacent internal capsule during the postoperative period of observation dating to three months. Microscopically, the laser lesion in the basal ganglia was cystic. Astrocytic proliferation was delayed for one to two weeks. Furthermore, hematoma formation or excessive bleeding did not occur at the level of 8 and 15 joules energy density. However, when 30 joules laser energy was delivered into the rod, hematoma formation was occasionally found and mortality rate was high. The internal capsule appeared normal or showed minimal gliosis, but the nerve fibers in the internal capsule appeared viable throughout the period of observation. Therefore, by use of appropriate light guides (such as the quartz rods), it is feasible to apply laser irradiation to specific loci within neural tissue. Furthermore, the possibility of clinical application is readily apparent because of two features of the laser effect in such regions, mainly, the absence of significant hemorrhage at appropriate energy levels and the apparent resistance of the internal capsule to significant damage by the laser.

FACTOR OF TISSUE PIGMENTATION IN LASER EFFECT ON NEURAL TISSUE

Goldman and others have shown previously that pigmentation in biologic tissues amplifies destruction caused by laser impacts from pulsed laser systems. It has been further observed that the blood filled tissues, such as angiomata of the skin, are more easily affected by laser radiation than normal skin. Our own results on studies of the nervous system demonstrated that laser radiation tended to concentrate its effects around the vascular tree and, in fact, appeared to be an important determining factor of the extensiveness of the lesion. Experiments have been performed to utilize this characteristic of laser action to even more precisely localize and enhance laser effects in neural tissue. Evans blue, a dye that remains confined to the vascular compartment, was administered intravenously into dogs and monkeys. Following this, craniectomy was performed and the exposed brain surface impacted with similar energy levels from focussed shots as utilized in experiments previously described whert dye was not employed. In addition, thalamotomy was performed after perfusion of such a dye. The resultant destruction from comparable energy levels was greatly magnified, presumably because the dye traversing in the vascular compartment at the moment of impact enhanced local absorption by laser. This phenomenon has been observed both with surface (cerebral cortex) and intracerebral (thalamotomy) laser impacts. When examined microscopically, such lesions were identical with other lesions except for the greater amount of tissue destruction. In one instance, a laser impact with 7 joules delivered into the basal ganglia induced a one centimeter cavity caused by vaporization of the tissues with necrosis. Thalamotomy performed on the other side of this animal prior to administration of the dye, utilizing 7 joules ruby laser radiation caused a discrete 4 mm area of necrosis, but no cavitation resulted. Because the amount of vascularity in gray matter is much greater than that of white matter, it is hoped to utilize this principle ultimately to even further confine the effects of laser irradiation to the desired area of brain tissue. Furthermore, if the use of such dye stuffs will permit application of extremely low levels of laser irradiation for such procedures, less expensive low output pulsed laser systems could be employed to advantage in similar research.

BASIC GOALS OF OUR CURRENT LASER NEUROSURGERY PROGRAM

As developed by McLaurin and Brown, the following neurosurgical program has been activated and is offered to serve as a model for current laser neurosurgery research.

1. *Investigation of different laser systems.* At the present time, the bulk of the work has been performed with the pulsed ruby laser systems and to a limited extent with the neodymium laser. Investigation must be expanded to intensive investigation of Q-switched versus normal mode systems and also with investigation of the second harmonic from these systems. It appears vital to search for laser irradiation of differing but specific wave lengths to provide information in order to attain those wave lengths that will produce the most desirable effects in different areas of the brain. The advent of recent high powered CW and quasi continuous lasers offers furthermore the promise of use by the neurosurgeon as a "bloodless knife" and studies have just been initiated in investigation of the latter.

2. *Neurophysiological and neuropathologic applications.*

a. Neurophysiology—To increase our knowledge of the function of discrete areas of the brain, the pulsed laser systems can be utilized to study the effects of localized and discrete lesions in various portions of the central nervous system with an accuracy heretofore impossible to achieve. Such studies will permit a fuller understanding of neurophysiological function and neuropathological disorders to enlarge our understanding of brain function.

b. Neuropathologic investigation—Investigations are planned to provide screening methods for use of drugs in cerebral edema and to study the pathophysiology of head injury. Such investigations will provide much needed information for the clinical corollary.

3. *Search for clinical applications.* It must be emphasized that clinical applications should be sought only after appropriate search to establish threshold levels for safety and maximal benefit to the patient undergoing treatment, and to assure the absence of long-term detrimental effects.

a. Thalamotomy—Experimental results have demonstrated the ability of appropriately transmitted laser beam (by quartz rods) to produce lesions which appear to offer distinct advantages over conventional means for this surgery in clinical therapy. However, more data is needed to determine the permanence of such lesions, to rule out harmful side effects, and to rigidly assess the possibility of the existence of brain damage in such subjects.

b. Selective anterior hypophysectomy—Investigation is also being undertaken to determine whether selective removal of the anterior lobe of the pituitary is possible with the laser, leaving the posterior lobe intact. Our experimental results indicate that the effects of laser impacts from a pulsed ruby laser directed to the pituitary are confined almost entirely to the anterior lobe of the pituitary. Apparently the laser beam has little effect upon the posterior pituitary which is predominately white matter. Since clinical hypophysectomy is accompanied by complications directly attributable to removal of the posterior pituitary, such as water and salt imbalances, it is desirable to perform selective removal of the anterior pituitary to avoid these complications.

c. Laser treatment of brain tumors

1) Direct impact. Studies have been performed by Rosomoff where the bed of primary brain tumors has been irradiated with small levels of laser energy. With utilization of higher power systems or with the cw laser beam, it appears possible that bloodless removal of these tumors could be effected.

2) A search is being made for selective tumor dyes, substances that will stain largely the neoplastic tissue to aid in localization of laser irradiation effects to only the abnormal tissues. Achievement of this goal will permit maximal preservation of adjacent and uninvolved normal brain tissue.

d. Laser transection (functional and/or anatomic) —The relatively high rate of occurrence of pain following operations to relieve tic douloureux, post-therapeutic neuralgia, etc., indicates the need for an improved method for treatment of these disorders. Nerve trunks have been shown experimentally to be interrupted with appropriate use of laser irradiation with permanent neurological sequalae, as expected, post-operatively. Since scar formation is minimal after laser irradiation, as shown by three-months follow-up observations, and considering further the possibility that recurrence of pain is related in some instances to growth of neuronoma at the end of the nerve trunk, perhaps laser irradiation can be effectively employed to avoid this complication and therefore improve the number of satisfactory results following such surgery. By tractomies by use of a very precise beam, such as the laser beam, it is feasible to consider interruptions of specific tracts in the spinal cord or brain stem for relief of pain or in the treatment of certain motor disorders. It would appear that the incidence of significant hemorrhage would be negligible following such impacts based upon experimental observations, and furthermore that the depth penetration of the laser could be precisely controlled. Application of a continuous wave beam of appropriate wave length to produce the desired results, would further permit desired depth of cutting under direct observation of the surgeon.

CONCLUSIONS

Investigations into the effect of laser radiation in neural tissues have disclosed that discrete and well-controlled damage of a desirable nature can be achieved. In highly selective and critical areas of the brain and spinal cord, the discreteness and precise control over such lesions appears desirable. Application of laser radiation as an aid to neurosurgical research appears well founded. Clinical applications, although not established, appear fruitful based upon observations from the present status of basic research. The initial results are encouraging and justify development of an extensive program to determine the future of laser as a device in clinical and experimental neurosurgical applications with the promise of improved methods at both levels in the future for the surgeon and the investigator.

REFERENCES

Astrom, K. E., E. Bell, H. T. Ballantine, Jr., and Heidenslebin: An experimental neuropathological study of the effects of high-frequency focused ultrasound on the brain of the cat. *J. Neuropath. & Exper. Neurol.,* 20: 484–520, 1961.
Brown, T. E., C. True, R. L. McLaurin, P. Hornby, and R. J. Rockwell: Laser radiation. I. Acute effects of laser radiation on cerebral cortex. *Neurology,* 16:730, 1966.
——— ——— ——— ——— ———: Craniocerebral trauma induced by laser radiation. *Life Sciences,* 5:81, 1966.
——— ——— ——— ——— ———: Laser radiation. II. Long term effects of laser radiation on certain intracranial structures. *Neurology, in press.*
——— ——— ——— and R. J. Rockwell: Thalamotomies caused by pulsed ruby laser. *Neurology, in press.*
Earle, K. M., S. Carpenter, U. Roessmann, M. A. Ross, J. R. Hayes, and E. Zeitler: Central nervous system effects of laser radiation. *Fed. Proc.,* Suppl., 14, (Part III) 24:129–139, 1965.
Fine, S. and E. Klein: Effects of pulsed laser irradiation of the forehead in mice. *Life Sciences,* 3:199–207, 1964.
Fox, J. L., J. R. Hayes, and M. N. Stein: The effects of laser radiation on intracranial structure. First Ann. Biomedical Laser Conf. of the Laser Medical Research Foundation, June, 1965.
Goldman, J. A. and R. Meyer: Transmission of laser beam through various transparent rods for biomedical applications. *Nature,* 205:892–894, 1965.
Goldman, L.: Dermatologic manifestations of laser radiation. *Fed. Proc.,* Suppl., 14, 24:S-92–S-93, 1965.
Liss, L. and R. Roppel: Histopathology of laser-produced lesions in cat brains. (Abstract) Presented at program of the American Academy of Neurology. April, 1965 (Cleveland).

Mendelson, J. A. and N. B. Ackerman: Study of biologically significant forces following laser irradiation. *Fed. Proc.,* Suppl., 14 (Part III), 24:111–115, 1965.

Rosomoff, H. L.: Effect of ruby laser on brain and neoplasm. Presented at First Ann. Biomedical Laser Conf. of the Laser Medical Research Foundation, June, 1965.

Stellar, S.: Effects of laser energy on brain and nerve tissues. *Laser Focus,* #15, 1:3–5, 1965.

————: Effects of laser energy on brain and nerve tissues. Presented at First Ann. Biomedical Laser Conf. of the Laser Medical Research Foundation, June 17–18, 1965.

17

Laser Treatment of Experimental Animal Cancer

The oncologist will find himself confused in experimental animal work with lasers unless he has training in laser technology and expert advice and suggestions. This has been the pattern today in cooperative studies of laser research of experimental cancer in animals as illustrated by the pioneering studies of McGuff and his group, by Klein and Fine and their associates, by Minton and Ketcham and by Rounds and Chamberlin. These exellent basic investigations both in laser radiation of tissue cultures of malignant cells and of experimental cancer in animals have shown varied patterns of effects of laser on cancer. According to Fine and Klein, the effect of the laser treatment of cancer in animals has extended ". . . from complete regression to accelerated deterioration of the tumor-bearing host." Again, these experiments show, as expected, that the color and consistencey of tissue affects the results.

MELANOMA

The melanoma is the experimental tumor of choice today in laser therapy. This can be the melanoma transplanted in the hamster cheek pouch and the mouse melanoma. The following melanomas in animals have been studied:

1. Cloudman S-91 melanoma
2. Harding Passey melanoma
3. Pitt 4 melanoma implant

The animals used have included hamsters and various mouse strains. One simple but important consideration in animal research is correct dosage of anesthesia. Ritter of our laboratory recommends the following, from Barnes and Eltherington, for anesthesia in milligrams of pentobarbital (Nembutal) per 1,000 grams body weight:

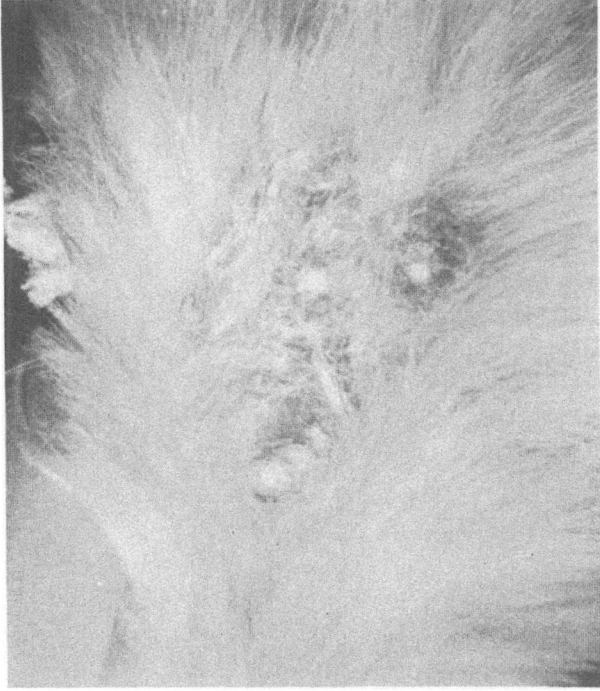

Fig. 17–1A. Papillomatous tumor in white mouse produced by painting with methylcholanthrene (Barich).

	Mouse	*Rat*	*Guinea Pig*	*Rabbit*
Intravenous	35	25	30	30
Intraperitoneal	60	50	35	40

Again, detailed controlled experiments are necessary including other types of therapy, including especially electro-surgery, thermal cautery, xenon light, x-ray grenz ray and the plasma torch. The absorption of the ruby or neodymium or argon laser produces a severe destruction in melanoma, at times spreading beyond the target area. When the laser impacts a lesion, the unfocussed beam will be approximately the size of the laser crystal diameter with the solid lasers. When this beam is focussed to a smaller area in order to provide more maximum energy density, the target area, then, of necessity, would be smaller in area. The advantage of focussing is that a higher energy density per unit area of the tissue is secured. It is difficult to give exact figures for M.E.D. (minimum energy density) required for complete destruction since there are many parameters for this. A tentative figure for adequate tissue destruction of melanoma of 2,500 joules/cm^2 is given for the pulsed ruby laser. In melanoma

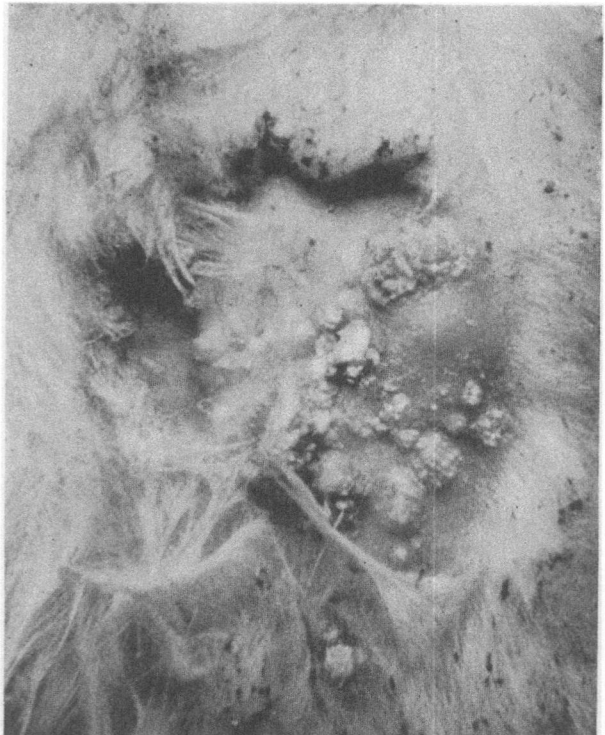

Fig. 17–1B. After laser impacts (685 joules/cm² energy density, unfocussed beam, pulsed ruby laser) showing only slight superficial charring reaction.

studies of McGuff there was found progressive necrosis from this target area. This progressive necrosis has been observed also by Helsper and his associates in experimental laser radiation of metastatic melanoma nodules in a patient. However, other investigators have reported that sometimes there is no progressive change and that the tissue destruction is limited only to the impact area. Our own clinical studies with the laser treatment of melanoma indicate that only occasionally is there progressive spread only to a limited area. The multifaceted characteristics of the laser impact indicate that progressive spread of tissue necrosis could continue through sonic wave and pressure change, spreading thrombosis and even immunobiologic reactions in tissue. On the other hand, viable plume fragments may develop new foci of melanomas about the target areas.

OTHER TUMORS

Many other animal tumors have been treated, including carcinomas, sarcomas, heptatomas and tumors induced by the polyoma virus. This

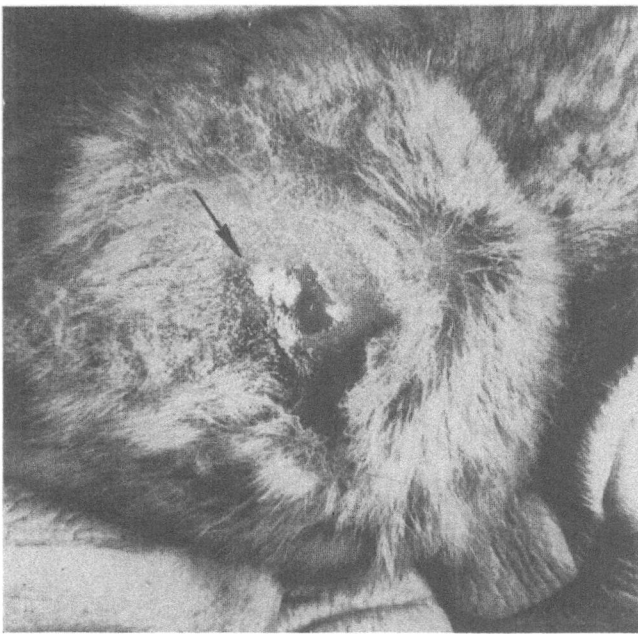

Fig. 17–2. Hamster with 4-day growth of polyoma virus tumor (developed by Sabin) after impact with pulsed ruby laser (reported to be 500 joules/cm² energy density).

list will increase. Our laboratory has worked with the epidermoid carcinomas in the white rat produced by methylcholanthrene painting by Barich, the polyoma virus tumor in the hamster provided by Sabin, the transplant of a malignant vascular tumor from thorium in mice produced by Swarm, and the C_3H/HEN mammary adenocarcinoma in rats for selective dye absorption by Brown and Ritter. Often, these results of laser treatment are not as spectacular as with the melanomas. This is true, as Minton has shown, if the tumors are hard and firm with much fibrosis. In our experiments with the hyperkeratotic or warty cancer of white rats, even high energy ruby lasers produced little superficial change. In our small series, subcutaneous tumors induced by polyoma virus responded to high energy laser treatment. However, even non-pigmented tumors, as McGuff has shown, have disappeared after laser treatment. Whether there are specific wavelength requirements as Minton indicates from his studies of wavelength absorption of tumor homogenates or whether there are non-specific effects from tremendous energy and power outputs from ruby, neodymium and argon lasers is not evident yet.

Little work has been done on the cancer immunobiology of laser treatment. Klein and Fine found it difficult to re-inocculate melanoma in animals after laser destruction of the melanoma. In cancer immunologic studies in our laboratory, in conjunction with laser research, Blaney is attempting to detect antibodies after laser impact. So far, this has not

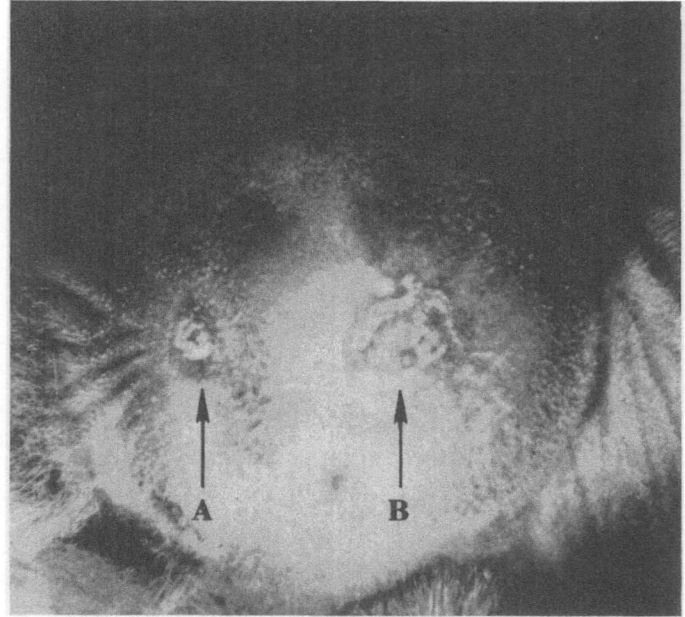

Fig. 17–3A. Hamster with SV_{40} virus ("Melnick") provided by Sabin. Impact 13 days after inoculation with pulsed neodymium laser *(Eastman Kodak)* , 500 joules/cm² energy density. Focal length of lens—240 mm. Spot size—⅛ in. at lens. Focussed 3 cm below surface. B. Same as 17–3A, but focussed 4 cm below surface.

been possible with fluorescent antibody techniques and tanned red cells. In hamsters with polyoma virus tumors treated unsuccessfully by lasers, Sabin found no antibodies after laser treatment.

CONTROL STUDIES

In brief, as usual, many controls are necessary especially for those in laser research who do not have much experience in animal cancer research. Some precautions include:

1. Know the natural untreated course of the tumor under investigation since some may disappear spontaneously;
2. Know the laser used and how to measure its output;
3. Have adequate numbers for critical statistical analysis;
4. Detailed records must be kept; these include pictures, gross and microscopic;
5. Have control therapies and try to approximate energy and power densities:
   a. Regular surgery
   b. Electro-surgery

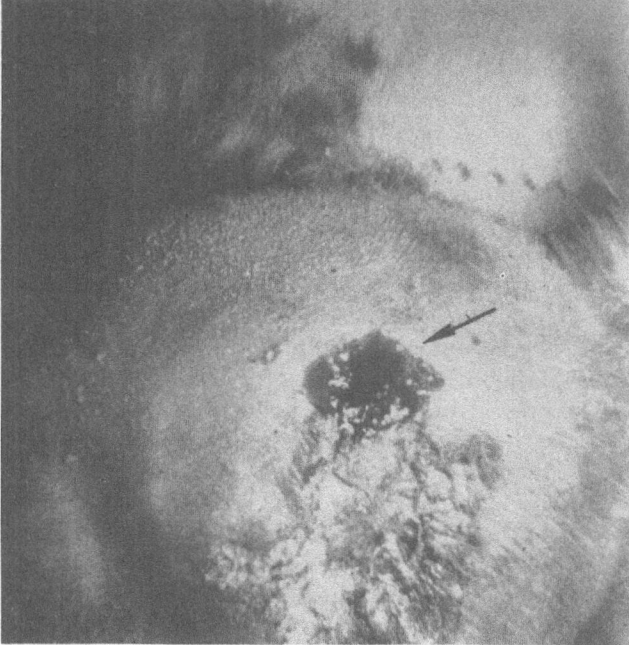

Fig. 17–4. Hamster tumor of live SV_{40} ("Eddy") virus, provided by Sabin. Impacted 20 days after inoculation with pulsed neodymium laser (*Eastman Kodak*), 510 joules/cm^2 energy density, unfocussed beams.

 c. Plasma torch
 d. Thermal cautery
 e. Other forms of radiation—xenon, x-ray, grenz ray and, where indicated, ultraviolet;
6. Try to evaluate critically and to determine if laser treatment caused spread of the tumors. Minton believes this danger is less with hard than with soft tumors; such spread is possible from plume fragments of the tumor;
7. Compare results with other investigators and try to arrange for cooperative studies;

 The nature of this tissue damage in tumors induced by the laser is receiving considerable attention by pathologists. It is defined, in brief, as a type of coagulation necrosis. This tissue necrosis is similar in some respects to thermal damage to tissue. Certainly, thermal damage is one feature of the effect of the laser on tissue. However, the tissue sections often show skip areas of destruction. This may suggest that the radiation is not homogenous, but it indicates also that there is selective absorption which does not occur with thermal treatments. At times in superficial tissues, resistance of the pilo-sebaceous unit is noted. This is of interest in Klein's observations of depigmentation of hairs of animals after laser impact. There may be greater destruction of hairs of animals after laser impact. Melanin absorption may not be the only question here. There may be greater destruction of blood vessels due to the absorption of the laser beam in the blood vessels. Secondary infections may occur in animals with far-advanced cancer even if they have not had laser treatment.

With high peak power outputs by Q-switching, ionization of air may be produced and morphologic evidences of ionization in tissue are different from those observed after laser radiation. Few studies have been done on laser impacts with prolonged pulse duration. The complete picture of the changes in tissue produced by high peak power outputs has not been fully worked out as yet, although the initial studies have been done by Fine, Maiman, Klein and Scott.

Concern is expressed by Fine and Klein that viable tumor fragments in the plume may be forced into adjacent tissue and vascular structures. We have not observed this mechanism following laser treatments of melanomas in patients. Inadequate cancer treatment of any type can be associated with continued spread and dissemination of the cancer. Viable cancer cells are found in circulating blood after cancer surgery and their significance is not clear. Viable fragments in the laser plume can cause air pollution of the laser laboratory and be a hazard to laboratory personnel. For the past three years we have used plume traps in laser treatments. There is no evidence at present of the carcinogenic effect of the laser in animals.

FOCUSSING IN DEPTH

It is possible with the lens system of the laser to focus in depth in the tumor and so cause tissue responses even at levels of 4 cm below the surface. The optics of this depth focusing have been discussed previously. These experiments can be done readily in the liver of animals and in large cancer masses. This is of interest in view of the experiments of Minton and Ketcham with the destructive powers of the laser in tumors of the liver of monkeys. It should be remembered that for focusing in depth, tissue destruction begins at the surface and goes down to focused depth in a cone shaped type of configuration, in melanoma nodules and in the liver. We have used this deep focusing technique also on large and deep cancers The energy densities deep in tissue are, of course, less than with focused impacts on the surface. However, with repetitive firings, sufficiently deep destruction can be obtained. The pigment masses in the superficial tissue may reduce the available energy density deep in tissue even more. In thermocouple expriements, Minton found below focus point of neodymium laser impacts in mice melanoma at temperatures of 100°C 1.5 cm below the tumor and 1,400°C 0.5 cm below an adenocarcinoma, a non-pigmented cancer.

Another method of getting the laser beam deep into tissue is with the special quartz rods previously described. This means puncture of the tumor or the incision of this mass to secure deep destruction in tissue. The rapid healing of liver necrosis induced by the laser is of interest. The varied patterns of this healing mechanism with laser impacts of normal liver and liver invaded by cancer are being studied.

With the argon gas of high output, this laser can function as an optical knife and remove successive layers of tissue and even in highly vascularized organs. If excision and destruction are complete, the tumor will not recur; otherwise it can do so. We have used this laser for the treatment of tumors in man also. In studies of healing after argon laser impacts, such parameters as output, duration of area contact and the use of tissue adhesives after argon laser surgery are being studied in our laboratory.

SYNERGISTIC EFFECTS OF LASER RADIATION

X-ray is one of the standard treatments of cancer in man and x-ray radiation has been used previously in animal cancer also. As indicated previously, Rounds and his associates have studied the potential synergistic effect of gamma radiation and laser in tissue culture experiments. This experiment of Rounds with tissue cultures shows the value of this technique in the study of cancer therapy. Rounds is attempting to determine the wavelength dependency of tissues and presumably this could be applied to cancer tissues also. We have taken tissue cultures of control and treated areas of cancer to study growth characteristics and karyotypes. Often the laser irradiated areas are without growth. In experimental adenocarcinoma, McGuff has reported that the tumoricidal effects of lasers and x-ray combined is greater than either alone. We have used combined x-ray and laser in squamous carcinoma in man, but cannot tell in uncontrolled experiments whether there is any value.

In human cancer, we have used cancer chemotherapeutic agents such as 5-fluorouracil and nitrogen mustards, combined with laser treatment. These chemical agents both have been used before and after laser treatment. It is difficult to evaluate results critically since the cytotoxic chemical agents, especially if applied or used topically, continue to produce rawness and destruction of the cancer. With Cloudman S-91 melanoma in CDF_1 female mice, Minton, Weiss and Zelen have used ruby laser with outputs of 48 to 56 joules and as the cancer chemotherapeutic cylcophosphamide. They believe that there is a greater effect of oncolysis or tumor destruction with the combination.

CONCLUSIONS

It is obvious that animal cancer provides an important medium for studies of the laser treatment of cancer. New instrumentation will continue to be assayed in animals. Again, the laser treatment of melanoma in animals produces the most striking results. Adequate areas of impact and what may be called loosely, "adequate" energy and power densities are required to destroy the cancer mass. Important basic studies must be continued as to wavelength requirements, factors of spread or lack of spread after laser impacts, more accurate definitions of the thermal, elastic recoil and pressure wave forces and electromagnetic field phases through better thermistor, transducer and ESR techniques. Studies on immunobiologic aspects of the laser treatment of animal cancer are just starting and shoud be continued. The carcinogenic effects of the laser and its possible spread of the cancer must also continue to be investigated in critical, and not purely speculative fashion. New treatment techniques with the argon laser, special flexible probes, quartz tubes and fibers for laser transmission for deep or inaccessible areas, exteriorization of organs for laser treatment and their replacement, synergism with other modalities of radiation and cancer chemotherapeutic agents all will be done in the near future. Again, control experiments with electrosurgery and other modalities of radiation should be done. Finally, it should be noted that one must be cautious in carrying over the data of the treatment of experimental or even spontaneous cancer in animals to that of man.

REFERENCES

Bach, J. L.: The amazing laser, medicine's newest research tool. *New Physician,* 13:178, 1964.

Blaney, D. J.: Personal communication.

Brown, T.: Personal communication.

Cromwell, N. H.: Chemical carcinogens, carcinogenesis and carcinostasis. *American Scientist,* 53:213, 1965.

Fine, S., T. H. Maiman, E. Klein, and R. E. Scott: Biological effects of high peak power radiation. *Life Sci.,* 3:209, 1964.

Goldman, L.: Laser cancer research. Springer-Verlag New York Inc., 1966.

Helsper, J. T., G. S. Sharp, H. F. Williams, and H. W. Fister: The biological effect of laser energy on human melanoma. *Cancer,* 17:1299, 1964.

Hoye, R. C. and J. P. Minton: The noble gas ion laser as a light knife. *Surgical Forum,* 16:92, 1965.

Klein, E., S. Fine, Y. Laor, M. S. Litwin, J. Donoghue, and L. Simpson: Laser irradiation of the skin. *Journ. Invest. Derm.,* 43:505, 1964.

———— ———— ———— L. Simpson, J. Ambrus, W. Richter, G. K. Smith, and C. Aaronson: Interaction of laser radiation with biologic systems: II Experimental tumors. *Fed. Proc.,* 24:143, 1965.

———— ———— R. E. Scott and S. Farber: *Proc. Am. Assoc. Cancer Res.,* 5:35, 1964.

McGuff, P.: Laser——A Surgical Tool. Exhibit A.M.A. June, 1965.

————: Comparative study of laser and x-ray radiation. 11th International Congress of Radiology, Rome, 1965.

———— E. Bushnell, H. Soroff, and R. Deterling, Jr.: Studies of surgical applications of laser. *Surg. For.*, 14:143, 1963.

———— R. Deterling, Jr., R. Gottlieb, D. Bushnell, F. Roeber, and H. Fahimi: Laser radiation of malignancies. *Ann. N.Y. Acad. Sci.*, 122:747, 1965.

———— ———— ———— H. Fahimi, D. Bushnell, and F. Roeber: The laser treatment of experimental malignant tumors. *Can. Med. Ass. J.*, 91:1089, 1964.

———— ———— ———— ———— ———— ————: Effects of laser radiation of tumor transplants. *Fed. Proc.*, 24:150, 1965.

Minton, J. P.: A method to determine laser wavelength absorption capabilities of experimental malignant tumors. *Life Sci.*, 3:1007, 1964.

————: Some factors affecting tumor response after laser radiation. *Fed. Proc.*, 24:155, 1965.

———— and A. S. Ketcham: The effect of ruby laser radiation on the Cloudman S-91 melanoma in the CDBA/2F$_1$ hybrid mouse. *Cancer*, 17:1305, 1964.

———— ————: The laser: A unique oncolytic entity. *Amer. J. Surg.*, 108:845, 1964.

———— ———— and J. R. Dearman: Tumoricidal factor in laser radiation. *Surg. For.*, 15:255, 1964.

———— ———— ————: The effect of neodymium laser radiation on two experimental malignant tumor systems. *Surg. Gyn. Obst.*, 120:481, 1965.

———— ———— ———— and W. B. McKnight: The application of pulsed high energy laser radiation to multiple intra-abdominal tumor implants in experimental animals. *Surgery*, 58:12, 1965.

———— C. D. Moody, J. R. Dearman, W. B. McKnight, and A. S. Ketcham: An evaluation of the physical response of malignant tumor implants to pulsed laser radiation. *Surgery* 121:538, 1965.

———— G. H. Weiss, and M. Zelen: Oncolysis with laser energy combined with chemotherapy. *Nature*, 207:140, 1965.

———— and M. Zelen: A method for predicting malignant tumor destruction by laser radiation. *J. Nat. Cancer Inst.*, 34:291, 1965.

———— ———— and A. S. Ketcham: Experimental results from exposure of Cloudman S-91 melanoma in the CDBA/2F$_1$ hybrid mouse to neodymium or ruby laser radiation. *Ann. N.Y. Acad. Sci.*, 122:758, 1956.

———— ———— ————: Some factors affecting tumor response after laser radiation. *Fed. Proc.*, 24:155, 1965.

Swarm, R. F.: Personal communication.

18

Laser Treatment of Cancer in Man

It is now time in the research program of investigations of the bio-medical applications of the laser to use the laser in the treatment of cancer in man. In brief, there are only two important questions to answer—is it necessary and is it safe?

NECESSITY OF LASER SURGERY

Surgical techniques in cancer have reached a plateau, and no significant advances have been made recently. The surgeon has turned to the engineer and the physicist for much needed help in the surgical treatment of cancer. The engineer and physicist have now given him laser surgery, cryosurgery, and recently the plasma torch. Are these developed enough for the research program? We believe laser surgery and cryosurgery are, but not yet the plasma torch. Now, in the discouraging field of cancer treatment, we believe there are areas where the laser can be used. In areas where conventional treatment can be used for cure and can be used more readily, the laser should not be used. This is so because, for the present, the cancer must be accessible to the laser or made so by the surgeon. High energy, expensive, complex laser instrumentation is necessary for laser cancer treatment and these new surgical tools are available only in a few medical centers. Also, should not the so-called "inoperable patient" be given a chance for laser treatment if laser can be used for his cancer?

SAFETY

We believe that in properly equipped medical centers with a definite program of area and personnel protection, high energy laser treatment is safe for the cancer patient. Is the laser itself carcinogenic?

168

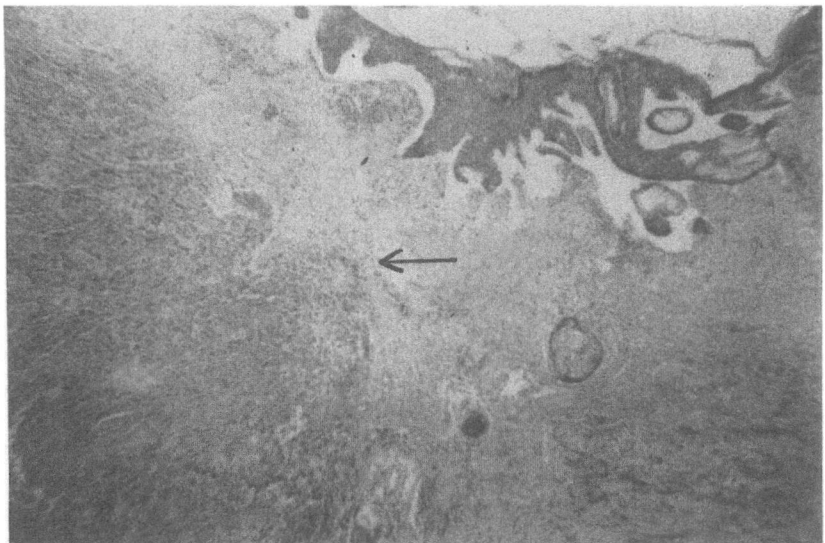

Fig. 18–1. Microscopic section after laser impact of melanoma showing sharply localized deep necrosis. Hematoxylin-eosin. ×360.

Ever since the early days of research in cancer with ionizing radiation, there has been interest and concern. These were the first environmental carcinogens to be investigated. From a theoretical aspect, the ionizing radiation may produce cancer itself or there may be a possible synergism between radiation carcinogenesis and chemical or hormonal carcinogenesis. Such studies have been done in detail with gamma radiation. As yet, there is but little data available in regard to laser radiation. The significant points for laser radiation are:

1. The tissue culture studies with the occasional development of chromosomal aberrations after laser impact (controls?).

2. ESR studies on tissues—lack of controls and only occasional changes.

3. The ionization of air produced by high peak power Q-switching. No evidence yet of ionization changes in tissue in a period of observation of only five years.

4. Occasional spread of animal cancer, but no evidence yet of human cancer. Factors may be mechanical and inadequate treatment rather than immunologic.

The histological changes, in both animal and human tissues, at least for the acute periods, do not resemble the changes which are induced by gamma radiation. There is a paucity of material available in follow-up studies of human tissues after laser radiation. Some of the material is of only five-year duration. In this material, there is no gross evidence of the development of any malignancy. In histopathologic studies, the changes are those of a non-specific fibrosis without any of the vascular changes or cellular changes usually associated with chronic radiation effects. Patients have had exit energies of over 200 joules ruby and neodymium lasers and 85 megawatts peak power output of Q-switched ruby laser on normal skin without malignant change. As yet, work with argon and carbon dioxide lasers is still only recent. We are studying the effects of chronic impacts in animals to determine the carcinogenic potential of laser radiation. Such studies should be done also on specific leukemia-susceptible strains of mice to see if leukemia develops and then to determine whether this leukemia is transmissible. As yet, there is no data available on the activation of leukemogenic viruses by laser radiation. Experiments with He-La cell cultures exposed to low energy radiation and then infected with herpes simplex are under way to determine whether such irradiated cells are more susceptible to virus infection.

The latent period, then, after laser radiation is still too short in the present scheme of things to determine the significance of this exposure. In the meantime, there is need for good controlled studies on this factor and similarly continued need for the protection of laboratory personnel from undue exposure. These include protection from exposure of tissue like superficial blood vessels, air pollution especially from plume fragments, as well as the significant protective studies on the eye. As mentioned, the hazards of the laser plume are of more concern to the operator than to the patient. We have no evidence as yet of laser-induced spread of cancer in man as in animals.

So as it is, we do not have any information now, that the laser is carcinogenic. Since we are, at present, treating local cancer, the patient should be protected from undue exposure. As indicated, our chief concern is eye protection, not fear of cancer. In brief, the patient who needs laser surgery should receive it, if it is available.

THE INSTRUMENTATION

High energy lasers are needed, for one cannot say laser treatment has failed unless one uses the best available. For the clinical investigative studies of the laser therapy of cancer of man, research cannot be limited to only one laser. Some lasers will be used to destroy the tumor in situ or block its progress. Some will be used to excise the lesions. With ruby lasers, exit energies up to at least 400 joules are needed in current clinical research programs. This we have obtained from the instrumentation of Maser Optics and now from Applied Lasers Division of Spacerays Inc. Often lower energies may be all that are required, especially in neurosurgery. Large metastatic areas, thick hard tumors, non-pigmented types—these are but some indications for high energy and power densities. For the neodymium, we have used an experimental

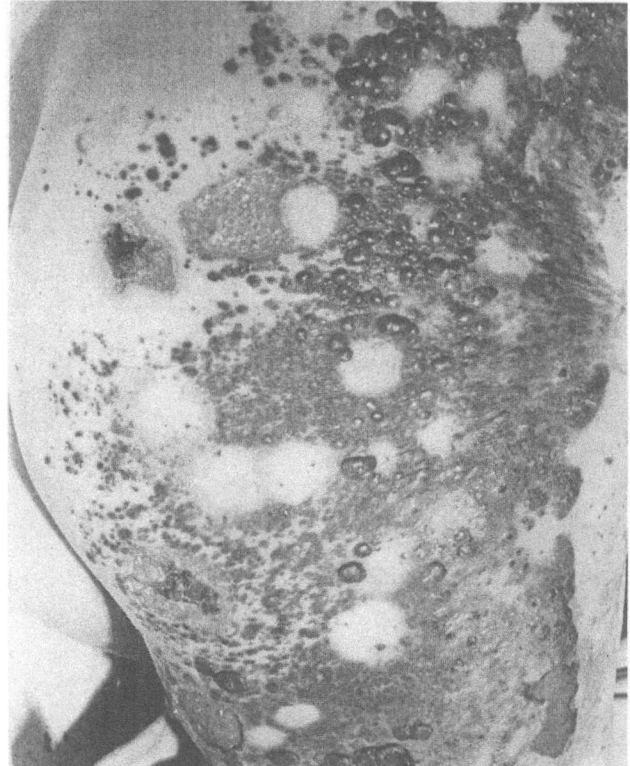

Fig. 18–2A. Metastatic cutaneous melanoma seven months after multiple impacts pulsed neodymium laser, experimental model, (*Eastman*) 656 joules/cm² energy density; showing persistent healed areas after surface-focussed and depth-focussed impacts. There was no aggravation of pre-existing extensive metastases about treatment areas. In addition, patient had spontaneous clearing and fading of many nodules. These local changes had no effect on visceral metastases.

laboratory model from Eastman Kodak with 1160 joules output. Special plume traps are necessary. For the argon laser, in our experience, 4 to 10 watts may be necessary. This we have done with the argon laser on loan from Bell Telephone Laboratories and now with the argon laser of Applied Lasers Division of Spacerays Inc. For the carbon dioxide laser (Perkin-Elmer), our current studies are with 10 to 25 watts. Ketcham at the National Cancer Institute has developed an extensive laser surgical operating room with an 800-joule neodymium laser. At present, in the United States, laser research surgical units are planned for Ohio State University, Johns Hopkins University, McGill University, Tulane University and Minnesota.

Mention has been made of the flexibility and safety of the laser for use

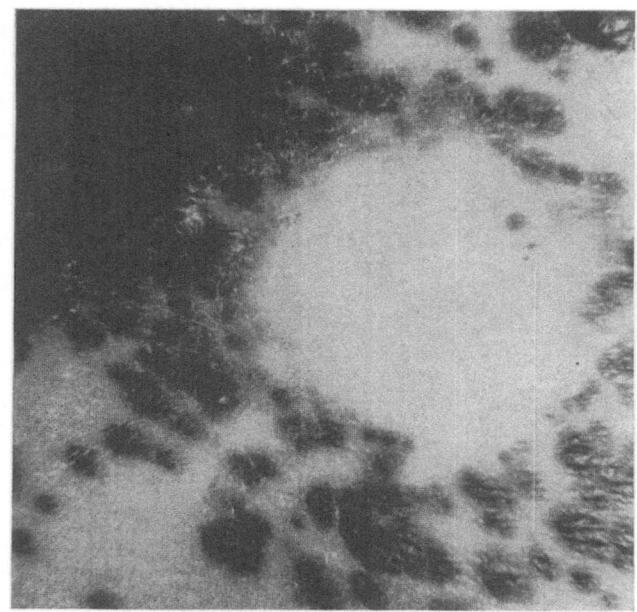

Fig. 18–2B. Showing details of cleared zone, biopsies were negative. No new satellite lesions observed after seven months.

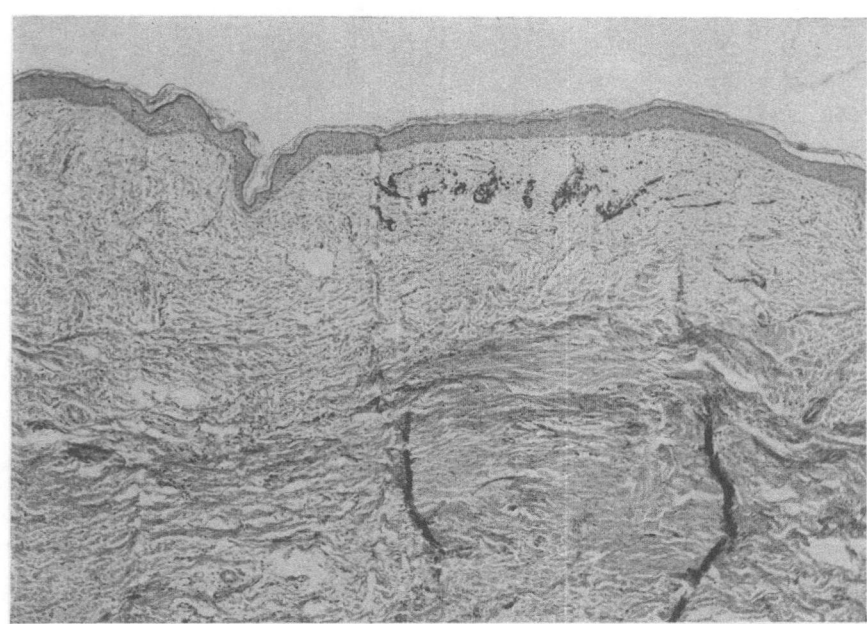

Fig. 18–2C. Biopsy studies at edge revealed melanin and phagocytes; no melanoma hematoxylin-eosin. ×400

Fig. 18–3. Showing various laser studies on patient with multiple subcutaneous melanoma nodules. Controls included electrosurgery. A. Incision and insertion of curved, tapered quartz rod for impact of exposed nodules. B. Incision of skin and excision of nodule by argon laser, 2-watt output cw experimental model, on loan from Bell Telephone Laboratories. C. Metastatic melanoma nodule excised by argon laser by Brown and Henderson. D. Character of tissue base remaining after excision—no hemorrhage during or after the operation. Closure by interrupted sutures. Lesion healed well.

E   F

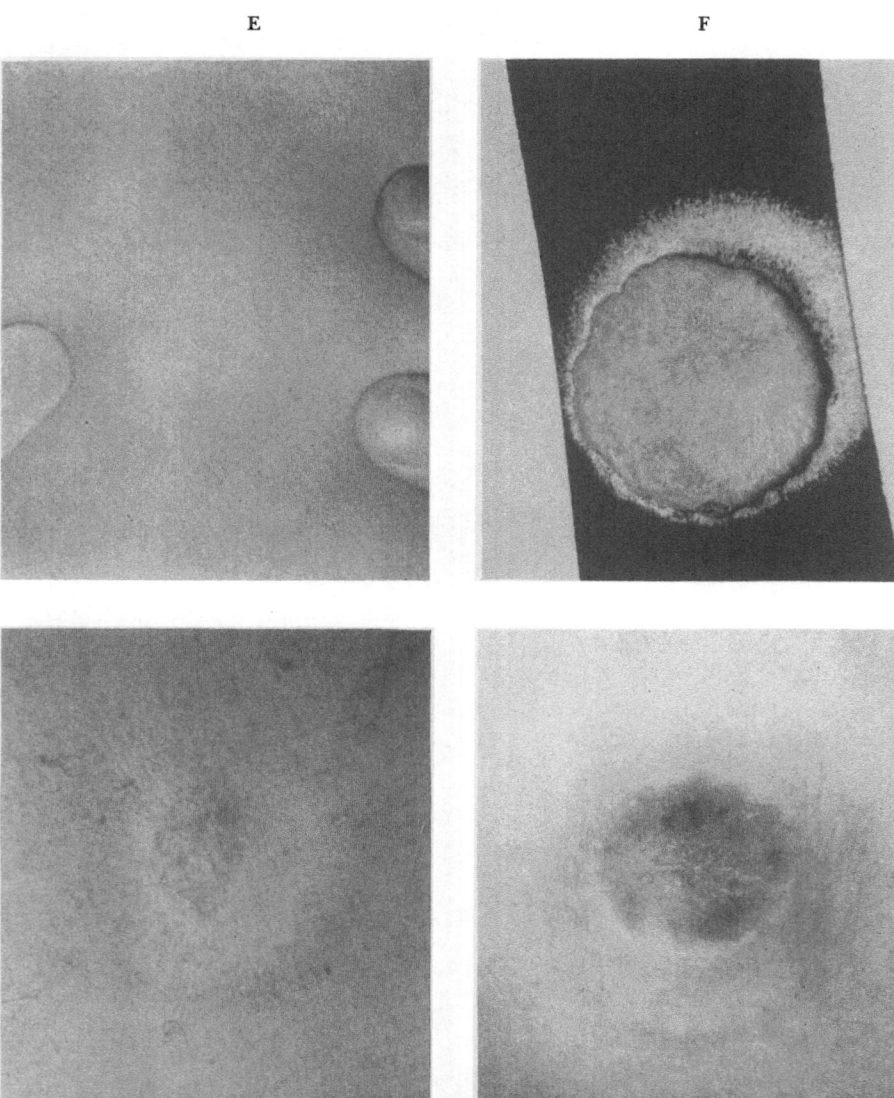

G   H

Fig. 18–3. E. Subcutaneous metastatic melanoma nodule. F. After impact over cardboard protectant, 340 joules/cm² energy density, unfocussed beam, pulsed neodymium
laser (*Eastman*). G. Cardboard off, showing extent of char. H. Four days later, showing
superficial necrosis, melanoma unaffected.

in the operating room, and these requirements will not be repeated. If the high output laser cannot go into the operating room, it will have but limited use in surgery. Our present laser operating room unit is a flexible ceiling-suspended unit under the direct control of the operating surgeon. This unit has been developed by Applied Lasers Division of Spacerays, Inc.

The difficulties of measurement have been mentioned and this is even more disturbing with the greater need of knowing the output of the high energy systems. The dosage for each patient must be recorded, as indicated before, in the following.

1. Laser—type and exit energy
2. Pulse—duration and character
3. Target area—character and size
4. Area density
5. Lens or rod transmission system
6. Target area—morphologic and physiologic characteristics

PROTECTION

The details will be listed in Chapter 20, but the essentials will be repeated here for emphasis. The focussing system of the laser should be intense enough so that the laser may be used in the usual well-lighted operating room. In such a room, eye protection can be more effective. The patient's eyes must be covered. For impacts about the face, we have used both protective glasses and cloths. Undue exposure of the target area can be prevented with cloths or protective cardboard. These can be sterilized. There also must be consideration of protection of adjacent structures such as blood vessels or nerves in the target areas. Operating personnel must wear eye protective goggles. If target areas must be held, arms and hands of the operator should be protected also with coverings or gloves. The special precautions with the use of argon and carbon dioxide lasers will be detailed also in Chapter 20. Only those personnel who have been trained in laser protection should be allowed in the laser operating areas.

CANCER TYPES FOR LASER SURGERY

In addition to reports of the treatment of cancer from our laboratory, a variety of cancer types have been reported by Helsper, Sharp, Williams and Pfister; McGuff, Deterling, Gottlieb, Fahimi, Bushnell and Roeber; and Rosomoff, Hellstrom, Brown and Carroll; and Kramesh and Phillips. The cancers includes melanomas, epitheliomas, cancers of the breast, lymphomas and sarcomas. All were accessible types. No matter what the cancer, the laser did cause destruction of the cancer, at least in the target area. Some of the patients with melanoma and epitheliomas whom we have followed have shown no evidence of cancer during the

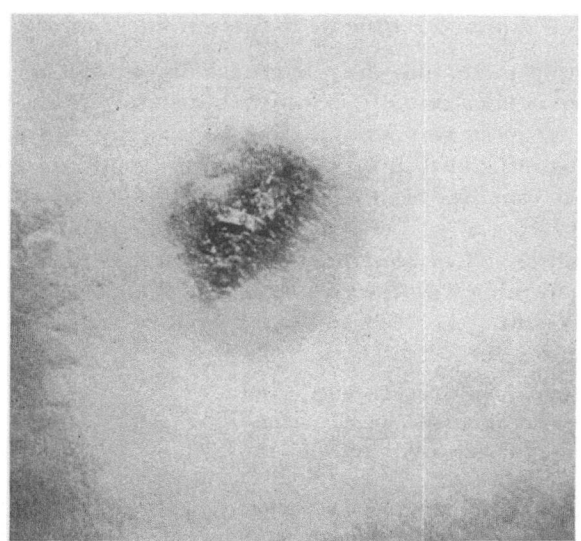

Fig. 18–4A. After multiple impacts, pulsed neodymium laser, 200-joules cm² energy density, basal cell cancer of upper back.

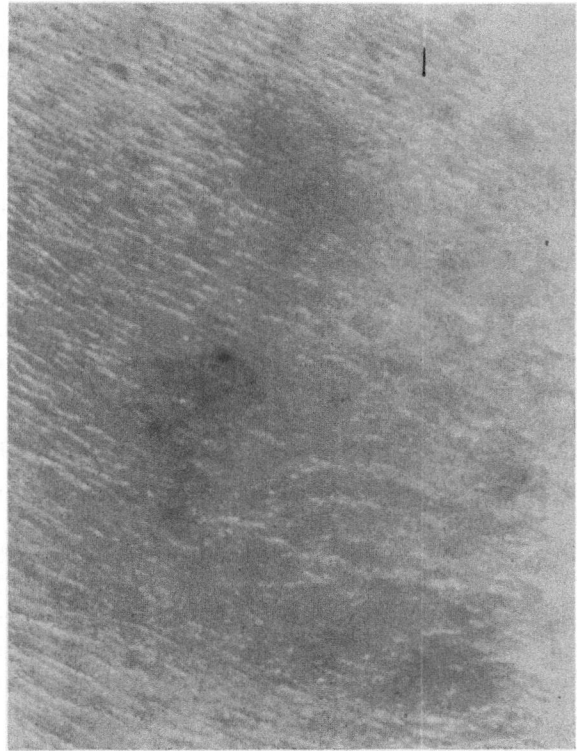

Fig. 18–4B. After eighteen months, showing only faint redness biopsy negative.

follow-up period, now over four years in some patients. These initial results, then were encouraging enough to warrant continued studies of the laser treatment of cancer.

The prime indication is for melanoma which is accessible and recurrent after treatment. With Dr. V. E. Siler, Professor of Surgery and Director of Laser Surgery in our laboratory, melanomas deep in the neck have been exposed and treated without any reactions. Also, deep recurrent melanomas of the thigh, leg and a large superficial melanoma of the face—3 by 3.5 cms—were treated. The melanoma of the face was negative on biopsy one year after laser treatment. The patient died later from a kidney infection and no evidence of melanoma was found at autopsy. The other melanomas treated have not spread. McGuff treated extensive metastases of melanoma with temporary disappearance of all lesions. McLaurin and Brown of our laboratory have treated recurrent melanoma of the skull adjacent to the dura. This was recurrent after removal of the involved part of the skull. The area was again removed and 80 laser impacts were given to the operative site through special quartz rods without damage to the brain. In addition to laser treatment, we have made extracts of the melanoma and given them to the patient after laser radiation. These were done as part of our investigative studies in cancer immunology. We have tried to re-inoculate in man melanoma tissue impacted by laser. In this study in one patient, controls included melanoma tissue treated with x-ray, and untreated tissue. There was no evident growth in any of these inoculated sites.

Multiple, accessible, superficial malignancies are the next indication for laser treatment since it is difficult to excise or cauterize skin areas with many malignancies. These multiple malignancies allow opportunity also of control therapies on other areas with other methods of treatment. We have treated also leukoplakia and superficial cancer of the lip.

Malignant tumors of the blood vessels are the next indication. These are called angiosarcoma and Kaposi hemorrhagic sarcoma. When the lesions are small and accessible, high energy laser is effective. With broad and extensive lesions, other forms of radiation or radioisotopes given internally should be used.

When cancer is inoperable, and if many of the lesions are accessible, then laser treatments should be offered. If the cancer is deep in tissue, as in the breast, bone, chest, or prostate, at present laser treatment cannot be used. If the surgeon can expose an area like the abdominal cavity and make the cancer accessible, then the laser treatment should be given to as many spots as possible.

CONTROLS

Since laser cancer surgery is still investigative, controls should be done especially when there are multiple cancers of the same type in the same patient. These controls, are, of course, excisional surgery, electro-surgery, thermal cautery, and other forms of radiation including xenon, x-ray grenz ray and the plasma torch. In this manner, with good follow-up studies, it will be possible to determine the real value of laser surgery in cancer.

MEASURES TO INCREASE ABSORPTION OF LASER IN MALIGNANT TISSUES

The natural coloring of hemoglobin and melanin have been mentioned previously in their relationship to the absorption of laser radiation. This was indicated above with the use of the laser in melanomas and vascular tumors.

Pigments to increase laser absorption have been described previously in Chapter 9. The dyes used may be either vital dyes or the cytotoxic dyes. Brown and Ritter of our laboratory are trying to produce synergistic effect of dyes with the laser in the treatment of experimental brain tumors with Nile blue and the laser. These dyes may be delivered to the tumor by direct injection, by intra-arterial injection, or even by intralymphatic injection. Most of the experience in laser cancer research has been by the direct injection of the dyes into tissue. This technique has very definite limitations in that the diffusion is fairly sharply limited to the area of the injection site. Experimental results indicate that the intra-arterial injection provides for a more effective distribution pattern of the dye in the tissue. The experience in perfusion techniques will help considerably in the use of dyes. There has been little data available as to whether the perfusion not only increases the concentration of dyes per unit area of the tumor, but also may make for more intranuclear absorption of the dye. For some non-pigmented basal cell cancers, we have used copper "forced" into the cancer by iontophoresis.

Recently, studies have been made on the intralymphatic injection of ferrite and then the radiation of the regional lymph nodes with electrical energy to generate heat of $50°C$ in tissue. Current investigative studies are utilizing this technique in animals to study the effect of laser impact on lymph nodes that have been suffused with ferrite particles. To increase laser absorption in metastatic lymph nodes, dyes may be injected or given into the lymphatics. Evans blue is used in these techniques.

Another technique of increasing the absorption of laser radiation has been the use of oxygen. In current investigative studies this has been of value in increasing the radiation effects of tumors. This can be done by various techniques such as a hyperbaric chamber and by the use of injections of hydrogen peroxoide. Investigative studies have been done in tissues of animals and man and then subjected to laser radiation. With the so-called polyatomic oxygen, hydrogen peroxide, and zinc peroxide we used in animals, there was no synergistic effect. The solutions of the so-called polyatomic oxygen are basically irritant to tissues on local injection and this interfered with controlled studies.

POSSIBLE SYNERGISTIC TREATMENTS WITH THE LASER

Obviously, the addition of any other form of treatment with the laser confuses the critical analysis of the follow-up program. Yet if we are doing investigative studies and if there is no evidence that we destroy all the cancer with the laser, we can add additional treatment in such diverse forms as:

1. Concomitant gamma ray radiation (described previously)
2. Autogenous tumor extract in various adjuvants
3. Cancer chemotherapeutic agents. We have used in patients topical 5-fluorouracil, cyclophosphamide, nitrogen mustard and methotrexate.

The details of such accessory therapies should be listed carefully so that at least an attempt at a critical analysis can be made. So far, this has not been possible to do in our patients, except with topical chemotherapeutic agents where we can use this on part of a cancer or on an adjacent, similar cancer.

THE TECHNIQUES FOR FOLLOWING THE PROGRESS OF LASER RADIATION OF CANCER

With the investigative nature of laser radiation at the present time, it is obvious that detailed follow-up studies are necessary. These follow-up studies should include not only the gross examination of the tumor and examination of the patient, but the use of frequent biopsies. To minimize the effect of repeated biopsy techniques on the progress of malignant lesions, fine small razor blade instruments have been used which give thin, deep sections. After the thin biopsy is obtained, then the malignancy can be closed by pressure and strips. Proper orientation by the technician of the thin bit of biopsy tissue is necessary.

Another biopsy technique which we have used is the chemocheck of Mohs. Here the tissue is fixed with the zinc chloride paste mixture and sections are made. This is used primarily to evaluate the extent of laser-induced coagulation necrosis in basal cell epitheliomas, three to four weeks after treatment. Excision of the laser-treated area with frozen sections was used also to determine whether additional laser therapy was needed. Unless follow-up biopsies are deep, recurrences under scars will be missed.

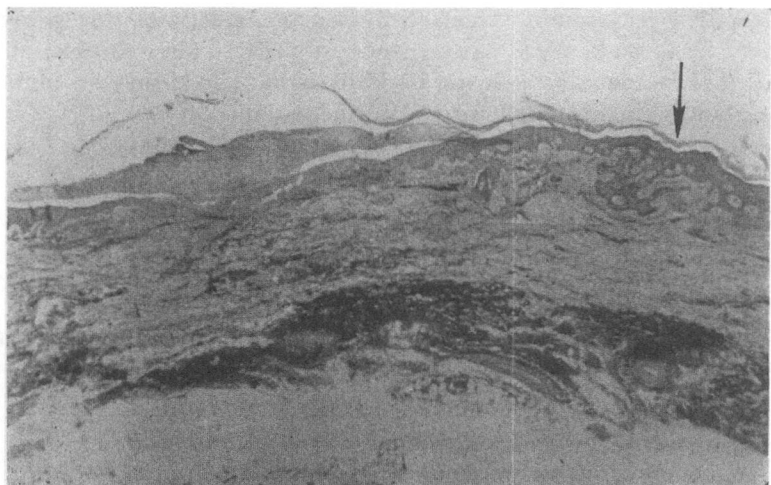

Fig. 18–5. Microscopic section after multiple laser impacts of basal cell carcinoma of forearm, showing area missed by the laser, hematoxylin-eosin ×25.

In addition to the gross examination and microscopic sections, efforts have been made to study the capillaroscopy of tumors to see whether there has been any change following laser radiation. This technique was developed originally by Urbach and is of importance as indicative of change in the activity of the tumor. This increased prominence of collateral circulation may be due to increased cellular mitosis, extension of tumor cords and the blockage of circulation. Frequent pictures with the standardized photographic techniques, as described in Chapter 6, should be used also.

With our Laboratory of Cancer Immunology, studies have been initiated on the immunologic aspects of laser radiation of malignancies of man. These studies include complement fixation using saline emulsion of the tumor, tanned cell hemagglutination, fluorescent antibody test, immuno-electrophoresis, etc. In addition, the serum of the patient treated, especially with melanoma, has been added to tissue cultures. At present, in cooperative studies with Siler and Blaney, we have been unable to detect any melanoma antibodies induced by the laser. The follow-up of the intra-abdominal lesions after laser radiation is a difficult job. X-ray studies may have to be done. Fiber optics instrumentation, especially with the peritoneoscope, may be employed in these selected areas.

So the patient following laser radiation deserves continued studies in an effort to find out the immediate and delayed effects of laser radiation in terms of the malignancy. The establishment of a central laser cancer registry and laser cancer commissions will assist in the collection of pertinent data, the most important of which is the frequency of recurrence of so-called adequately treated local lesions.

CONCLUSIONS

At present, there are limited but definite indications for laser cancer surgery in man. These include, in brief, accessible lesions which cannot be treated with conventional means. The treatment is investigative and should be controlled whenever possible by other modalities of therapy, especially electro-surgery. Follow-up studies are necessary.

The current problems of laser cancer treatments in man are not only of the hazards of the laser, but of the efficacy of this new, expensive and sophisticated instrument. Improvement in the development of high output lasers, especially the argon, carbon dioxide, and ultra-violet types, which are more flexible and can be used safely in the operating room and are able to be used by the surgeon—all this will assure more extensive use of the laser in cancer surgery. In this manner, laser cancer surgery in the future may be used for effective treatment of early cancer, not as now for the hitherto inoperable case.

REFERENCES

Ardenne, M.: Hyperthermia of cancer cells. *Med. World News* Sept. 24, 1965, p. 60.

Goldman, L.: The laser in medical research and surgery. *The New Scientist,* 21:284, 1964.

———— D. J. Blaney, A. Freemond, and P. Hornby: The biomedical aspects of lasers. *JAMA,* **188:**302, 1964.

———— J. Ingelman, and D. Richfield: Impact of the laser on nevi and melanomas. *Arch. Derm.,* **90:**71, 1964.

———— and R. Wilson: Treatment of basal cell epithelioma by laser radiation. *JAMA,* **188:**773, 1964.

———— ———— P. Hornby, and R. Meyer: Laser radiation of malignancy in man. *Cancer,* 18:533, 1965.

————: The response of skin cancer to topical therapy with 5-fluorouracil. *Cancer Chemotherapy,* 28:49, 1963.

————: Laser treatment of cancer. *Progress in Clinical Cancer.* Irving Ariel, editor. *In press.*

————: Laser cancer research. *Recent Developments in Cancer Research.* Springer-Verlag, New York Inc., 1966.

————: Experiences with the treatment of patients with high energy ruby and neodymium lasers. NEREM meeting. Boston, 1965.

————: Applications of the laser beam in cancer biology. *Int. J. Cancer,* 1:309, 1966.

———— V. E. Siler and D. J. Blaney: Laser surgery of melanomas. *Surg. Gyn. Obst.,* (Jan.) 1967.

Helsper, J. T., G. S. Sharp, H. F. Williams, and H. W. Fister: The biological effect of laser energy on human melanoma. *Cancer,* 17:1299, 1964.

Kellum, R. E.: Laser energy and malignancy—a review. *Arch. Derm.,* 92:734, 1965.

Ketcham, A. S.: Panel discussion on future biologic and clinical uses of the laser. The clinical point of view. *Fed. Proc.,* 24:172, 1965.

———— and J. Minton: Laser radiation as a clinical tool in cancer therapy. *Fed. Proc.,* 24:159, 1965.

Lasers and cancer. *Brit. Med. J.,* 5442:1080, 1965.

McGuff, P. E., R. A. Deterling, L. S. Gottlieb, H. D. Takimi, and D. Bushnell: Surgical applications of the laser. *Ann. Surg.,* 160:765, 1964.

———— ———— ———— D. Bushnell, and F. Roeber: Laser radiation of malignancies. *Ann. N. Y. Acad. Sci.,* 122:4747, 1965.

———— ———— ———— H. E. Fahimi, D. Bushnell, and F. Roeber: The laser treatment of experimental malignant tumors. *Canad. Med. Ass. J.,* 91:1089, 1964.

Meyer, R.: Personal communication.

Rosomoff, H. L., R. Hellstrom, J. Brown, and F. Caroll: Effect of laser on carcinoma in man. *JAMA,* 192:176, 1965.

Urbach, F.: Personal communication.

19

Dental Applications of the Laser

The teeth and oral cavity have been included in the study of tissues accessible to the laser beam. Although the possibilities of the development of laser dentistry appear to us to be excellent, there has been too little interest in the clinical and applied phases of laser dentistry by dentists and dental research groups. Research in laser dentistry by relatively few investigators has included the laser spectroscopy, laser impacts of teeth and oral cavity of experimental animals, impacts of extracted teeth of man, and rarely, the actual impacts of teeth in patients.

LASER INSTRUMENTATION

All varieties of laser instrumentation have been used by our laboratory in investigative studies of laser impacts on teeth. These have included pulsed and Q-switched ruby laser, neodymium lasers carbon dioxide and argon lasers. As yet with outputs of 2 watts, argon lasers have shown little change in the enamel. For impacts of the teeth in patients we have used special cuved quartz rods making for precise impacts on selected areas of the tooth. With prolonged exposure (2–3 minutes, 20–25 watt output carbon dioxide laser) we have reduced extracted teeth down to charred black fragments. Similar experiments have been done with the plasma torch. In addition, the laser microprobe has been used for spectroscopy studies of various portions of the teeth.

PROTECTION

The same principles of area and personnel protection apply to research in laser dentistry. The bright surface of enamel can produce strong reflectance. This may be of significance as potential hazards to adjacent structures in the oral cavity. As yet, closed laser systems of high

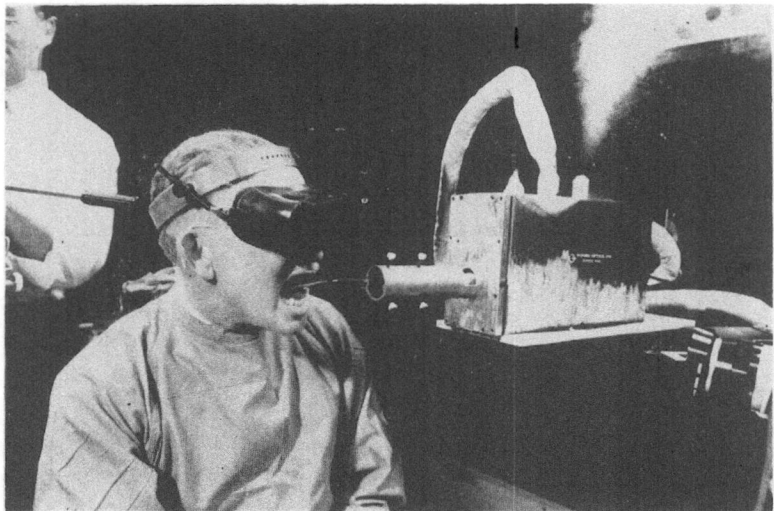

Fig. 19–1. Showing use of tapered curved quartz rod for impacting tooth, using pulsed ruby laser, liquid nitrogen-cooled, 20 joules/cm² energy density.

output are not available. However, as we reported in 1964, special curved quartz rods may help to localize the impact area and to avoid severe reactions to adjacent gums, tongue or cheeks.

The hazards of exposure problems to the laser plume have been considered before. The oral cavity is only a partially open cavity during laser impact and portions of the plume may be inhaled, digested as well as deposited on adjacent structures. Our studies with color infrared photography of the laser plume on tooth impact in the oral cavity of man show the plume to be localized to the target area. The laser microprobe does help in the analysis of the potential hazards of the laser plumes from laser impacts of normal and abnormal teeth. Our studies with thermistors placed adjacent to that tooth selected for laser treatment show but little elevation (10°C) after laser impact. In their studies of 35 and 55 exit energy ruby laser impact of the mandibular incisor of Syrian hamsters, Taylor, Shklar, and Roeber found some changes in the enamel of adjacent mandibular and maxillary incisors. As yet, there is no data available on the use of small barium titanate transducers to detect sonic wave reactions in structures adjacent to the target area. Laser transillumination problems about the face and neck have been mentioned previously. The only data available to us is that obtained from special photographic studies of the impact of teeth in volunteers.

Eye protection is used both for the patient and for the operator. On the patient, this would reduce the hazards to the eye from external sources of the laser. The potential hazards to the eye from transillumination through soft tissue and bones into the orbit are not known. After a series of teeth impacts repeated at intervals in a volunteer, no eye damage was observed.

The dental operator must wear protective glasses. With the use of the quartz rods for transmission of the beam, accurate placement of the beam is assured even with the use of protective glasses. Since laser surgeons must wear glasses for much more delicate laser operations, dentists should not object to wearing of these glasses until the other measures of eye protection are developed. Again, since the optical properties of the oral cavity is different in man than in animals, many of the actual and potential laser hazards will have to be determined by actual investigative studies in man. This will develop the field of oral laser surgery either for dentistry or for cancer surgery.

LASER SPECTROSCOPY OF THE TEETH

Laser spectroscopy has shown the value for the analysis of the various structures of the teeth. These have included enamel, dentin, carious dentin, cementum, supragingival and subgingival calculus. These examinations have been done by Sherman, Ruben, Goldman, and Brech. Some interesting findings included the unsuspected presence of titanium and zinc in calculi. It is hoped that laser spectroscopy will be an important technique for studies of the resistance of enamel, the changes of the teeth with varied types of cleansing routine, local infections and changes in the tooth in systemic disturbances. In addition those changes in the teeth which can occur with age can be studied also with spectroscopy.

ENAMEL

The highly reflectant surface of enamel has shown some measure of resistance to laser impact. Even with high energy lasers a small crater is formed. Prepared sections of teeth and x-ray photographs of these prepared sections have shown at time rather limited penetration of the enamel. The exact natures of these fusion changes in the impact area are still not determined. Sogannes and Stern, pioneers in laser dentistry, have shown that laser beam "glazing" of human dental enamel exhibits more resistance to vitro demineralizing solution than control teeth. Whether this would hold for increased resistance to caries, also, is not known. Kindersly, Jarabak, Phatak, and DeMent believe from studying teeth sections that larger holes were produced and in chalky enamel (early caries) than in non-chalky. Enamel stained by Evans blue and India ink in our laboratory has shown deeper craters. Q-switching in our laboratory with peak power outputs of 125 megawatts drilled small craters, but did not penetrate into the pulp cavity.

DENTIN

With the penetration of enamel, the dentin is also involved. A larger crater and charring may be formed. Peeling of the cementum from the dentin has also been observed. In some human adult teeth impacted on the optical bench with an energy density of 13,400 joules/cm² we were unable to detect changes in the dentin.

PULP CAVITY

An important structure to be considered in relation to laser impacts in the pulp cavity. If laser impacts on enamel dentin and on carious areas do not cause necrosis in the pulp cavity, then that fact, if true, would advance greatly the progress of laser dentistry. Involvement of the pulp cavity can be detected by the use of small thermistors placed inside the extract tooth and by microscopic sections of the tooth after laser impact. The thermistor probe is a crude detector of the effect of laser impact since the duration of impact is in terms of millisecond. In one extracted tooth experiment we detected a rise of 29°C in the pulp cavity after laser impact on the labial surface of the tooth. Taylor, Shklar, and Roeber found changes in pulp cavities of irradiated incisors of hamsters and some changes in pulp cavity of teeth distant from the irradiated teeth. In one volunteer subjected to repeated laser impacts as reported, no involvement of the pulp cavity was found. This problem of pulp cavity involvement can be settled in many only with studies carefully controlled of impact area of thermistor and transducer instrumentation and detailed microscopic and electron microscopic studies.

DENTAL CARIES AND CALCULUS

Darkish areas of caries have shown significant changes from laser impact. These were done in our laboratory three years ago and reported in 1964. Yet there has been little work in caries save for laser spectroscopy. At times, laser impacts have destroyed relatively large areas of caries. The efficiency of this treatment can be decided only by controlled histological sections to determine the depth of necrosis, the presence or absence of organisms. In some of the sections of the teeth we studied it was not possible to define the true extent of laser change. Again, damage and hazards to the pulp cavity will be determined by subsequent continued, controlled observations and detailed studies. If it can be shown that there is efficient and effective destruction of dental caries so that the tooth is ready for filling, this would be an advance in laser dentistry.

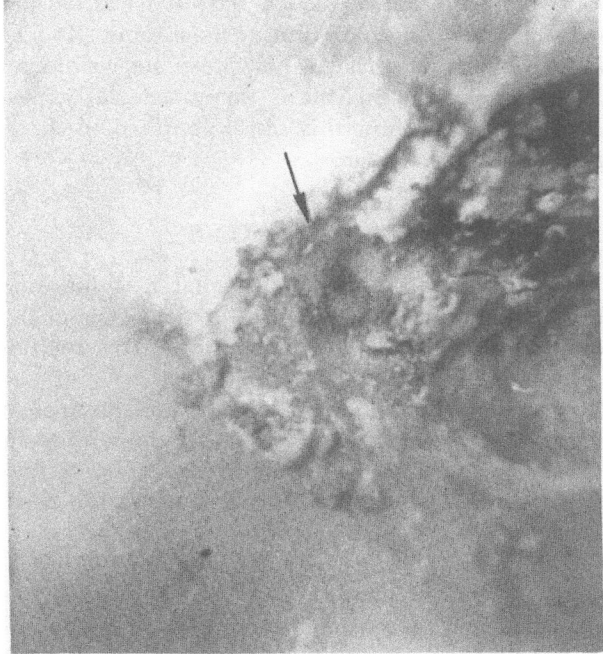

Fig. 19–2. Impact on pigmented carious area of tooth, 13,400 joules/cm^2.

As yet, we have been unable with the laser to fuse fillings and amalgams so that these could fill this laser treated area. With the temperature required perhaps it is of significance that this cannot be done.

Studies on impacts of tooth calculus or tartar have shown changes. Destruction of the calculus is produced. This destruction is increased even to complete vaporization. When the tartar is colored, either naturally or colored by dyes or by copper sulfate, laser destruction is extensive. The problem of selective destruction of calculus without damage to the underlying gum margin is a real one. This can be done only by controlled studies on patients.

LASER TRANSILLUMINATION

Laser transillumination techniques mentioned previously have been used also for teeth. Kinersly, Jarabak, Pathak and De Ment have demonstrated the value of this procedure for the teeth. In their instrument, a laser beam was transmitted through conduits, metal, rubber tubing and even hypodermic needles. Laser outputs varied from 0.1 to 0.5 joule. Beam diameter was 2 mm. Some teeth had holes drilled in them to bring light to the center of the crown. In the usual technique of transillumination, the room where the optical bench was used was darkened and the transillumination shown on color film (Ektachrome -X). The films revealed cracks in the enamel surface which were not visible grossly. These preliminary experiments reveal this is the value and experiments will be continued to determine how this can be applied to the examination of the teeth in the mouth of man.

CONCLUSIONS

These studies at present then indicate that a significant portion of the laser laboratory should be devoted to the field of laser dentistry. Unlike many dentists, we feel that this a profitable area for research, especially in the treatment of caries and perhaps even of calculus. The dentist and especially dental histopathologist and electron microscopist must work with the biologists and the physicians and the engineers engaged in laser research. The purpose of this cooperative study is to develop flexible, effective and safe laser instrumentation needed for laser dentistry. Dentists should be active in this program, not wait until other disciplines do the work for them. Even in investigative work on enamel and dentin, etc., the laser should join the other instrumentation of histology, histochemistry, electron microscopy and radioisotopes.

REFERENCES

Basil, B. G.: Future research on dental caries. *Ann. N.Y. Acad. Sci.*, 131:922, 1965.

DeMent, J., J. Jarabak, N. M. Phatak, and T. Kinersly: Profilmetry studies of laser impacted tooth enamel. First Annual Biomedical Laser Conference, Boston, June, 1965.

Goldman, H. M., M. P. Ruben, and D. Sherman: The application of laser spectroscopy for the qualitative and quantitative analysis for the inorganic components of calcified tissues. *Oral Surg.*, 17:102, 1964.

Goldman, L., J. A. Gray, B. Goldman, and R. Meyer: Effect of laser beam impacts on teeth. *The Journ. Amer. Dental Ass.*, 70:601, 1965.

————, P. Hornby, R. Meyer, and B. Goldman: Impact of the laser on dental caries. *Nature*, 203:417, 1964.

Gordon, T. E.: Laser effects on oral and other tissues. First Annual Biomedical Laser Conference, Boston, June, 1965.

Jenkins, G. N.: The chemistry of the placque. *Ann. N.Y. Acad. Sci.*, 131:789, 1965.

Kinersly, T., J. M. Jarabak, N. M. Phatak, and J. DeMent. Laser effects on tissue and material related to dentistry. *J. Amer. Dent. Ass'n.*, 70:593, 1965.

————, ————, ————, ————: Effects of lasers on teeth, an appraisal. First Annual Biomedical Laser Conference, Boston, June, 1965.

————, ————, ————, ————: Tooth transillumination with laser radiation. First Annual Biomedical Laser Conference. Boston, 1965.

————, ————, ————, ————: Laser induced microperforations in teeth sections. First Annual Biomedical Laser Conference, Boston, June, 1965.

Lasers and caries. *Lancet,* 2:949, 1964.

McGuff, P.: Laser's potential in the management of oral disease. First Annual Biomedical Laser Conference, Boston, June, 1965.

Priester, E. S.: Laser: Possibilities of its use in medicine and dentistry. *DDZ,* 19:120, 1965.

Rosan, R. C., F. Brech, and D. Glick: Current problems in laser microprobe analysis. *Fed. Proc.,* 14:127, 1965.

Schulte, W., et al.: Laser effects on dental substances—microscopic study. *Deutsch Zahnaerztl. Z* 20:289, 1964.

Sogannes, R. F.: Introduction to the problem of caries. *Ann. N.Y. Acad. Sci.,* 133: 1965.

———— and Stern, R. H.: Laser effect on resistance of human dental enamel to demineralization in vitro. *Journ. So. Calif. Dent. Ass'n.,* 8:32, 1965.

Stern, R. H. and R. F. Sogannes: Laser beam effect on dental hard tissues. *Interna'l Association Dental Research,* 43:873, 1965.

————: Personal communication.

20

Area and Personnel Protection From the Laser

In all phases of laser research, operating personnel must develop programs for protection against the laser beam. Safety measures must be planned in advance, adhered to and reviewed critically at frequent intervals. This program must be developed whether exposure to the laser is only occasional or frequent. Safety measures are also to be followed by patients receiving laser treatment and by visitors.

Although most biomedical research on lasers is done with the ruby laser, the neodymium laser and the argon laser, the protection programs must be flexible enough to include lasers of many different wave lengths and different energy and power outputs. Exit energies now may go to thousands of joules, with energy densities in the 100,000 joules/cm² range. Lasers operated in the Q-switched mode can now generate power outputs as high as gigawatts. With the generation of second and fourth harmonics, laser generation in the ultraviolet range is now possible. Outputs of 200 watts for CW argon lasers are also now possible. High output ultraviolet lasers are now available. As yet, no one knows the limits of the very dangerous carbon dioxide laser with outputs now more than 1,000 watts. Ionization in air and in materials has been suspected after impacts from high peak power lasers. Therefore, all these phases of protection must be considered.

In brief, the protection program centers about the following.

1. Personnel control
   a. Eye
   b. Exposed skin
   c. Inhalation
2. Area control
   a. Avoidance of spectral reflectance
   b. Proper ventilation
   c. Avoidance of electrical shock

A number of laser groups have devised effective and detailed laser safety manuals. Some of these current manuals include:

1. Graham W. Flint, "Determination of Laser Hazards." Martin-Marietta Corporation, U.S.A.
2. "Laser Safety." Martin-Marietta Corporation, U.S.A.
3. F. X. Warden, W. C. Roberts and J. P. Dunn, "Guidelines for the Safe Use of Lasers." Western Electric Co., U.S.A.
4. "Lasers and Masers—Standard Health and Safety Practice." Raytheon, U.S.A.

In addition, military services have their own respective manuals. An example is "Laser Systems—Code of Practice," prepared by the Chief Safety Officer of the Ministry of Aviation of Great Britain.

It is, of course, good industrial hygiene practice to have such a manual no matter how small or large the laser installation may be. It is necessary to have someone responsible for laser safety no matter how small the organization may be. It is also necessary to keep abreast of information available and to review your own laser safety program from time to time. Good training, discipline, and supervision of laser personnel will reduce the hazard and will also do away with the hysterical utterances and loose information regarding both the early and late hazards of the laser. In all phases of radiobiology, at least in the early phase, physicists, as indicated, are notoriously careless. Too little, as yet, is known about the late phases of the laser reaction in tissue. Some sceptics believe that even at this date, too little is known about the parameters of the laser reaction to define in precise quantitative terms that which is called minimal reactive dose (MRD) for such important organs of man as the eyes and the skin, of the ruby laser—even normal and, of course, Q-switched modes, the same for the neodymium, and all the gas lasers— argon, carbon dioxide, ultraviolet, and high output helium-neon. At this time, it is doubtful whether high output carbon dioxide lasers will be used for laser surgery in man. There is no doubt, however, that they will be used for industrial and military applications and much will have to be learned about the hazards for man of this invisible beam of 106,000 Å. Although the carbon dioxide laser will be used chiefly in animal experiments for vascular surgery and for surgery of the liver, spleen, and lungs, we have used this under nerve block anesthesia in metastatic melanoma of the skin.

AREA CONTROL

Of course, it is not always possible to separate rigidly personnel and area controls. But the distinction is made to emphasize the need for the development of special laser laboratories, not just the utilization of

Fig. 20–1. Soft frame welder type preferred to spectacle frame type.

laser equipment in any available laboratory space. The details of such special laser facilities, both for the laboratory and the operating room, will be listed in another chapter. In our special laser laboratories for patient treatment and research, the rooms are specially designed for area and personnel protection. Reflectance is prevented by the use of black colors. Warning systems are set up with lights and buzzers during the charging and firing of the laser. During the charging period, no personnel may enter the room. Proper ventilation is also provided, against air pollution from plume particles and from Raman and Brillouin experiments. Warning signs, designed by DeMent and modified by us, are used in all areas where there are lasers. The laser danger sign proposed by the Laser Systems Center of Lear Siegler, Inc., is particularly effective.

EYE PROTECTION

The first concern in laser protection continues to be the eye. The protection of the eyes means protection not only from impact of the direct beam but also from the significant amounts of reflection from surfaces. Eye protection is divided essentially into the following phases.

1. The design of laser systems to develop as much as possible the closed system technique, use of an electronic image converter tube or closed circuit television viewing.
2. The avoidance, as far as possible, of highly reflectant surfaces on and adjacent to the target areas.
3. Protection in the form of protective devices for the eyes.
4. Constant reappraisal of all proposals for area and personnel protection.
5. Continued eye examinations and records for all operating personnel, especially for the effects of long-term exposure.
6. Immediate examination of the eyes in any accident or suspected accident.

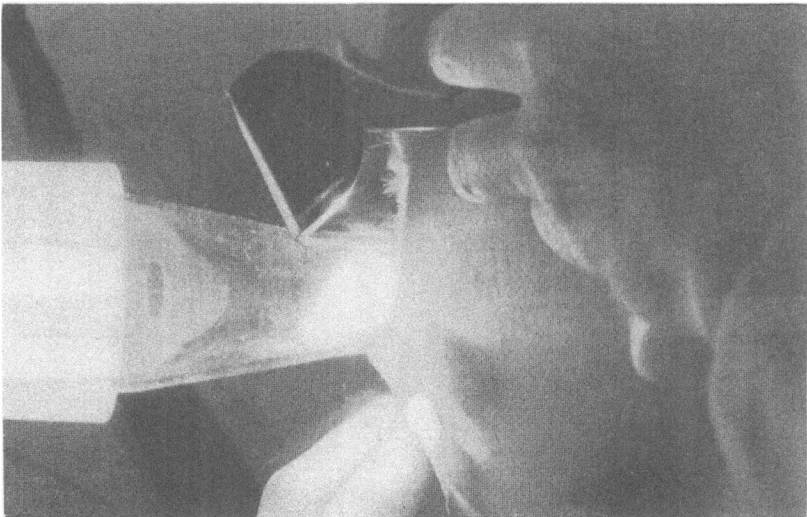

Fig. 20–2. Laser impact of face with cone resting on cheek showing reflection of beam under the spectacle type of protective glass.

Since laser instrumentation production is still not assembly line, investigative models will continue to be used, both in the research laboratory and in pilot experiments. Many of these instruments will be open with the laser beam freely exposed on its course from the laser head to the target area. Area control means the development of dark, dull, non-reflecting surfaces in the laboratory, and the use of extensive, heavy felt drapes and the use of devices to attempt to focus on the target area and to observe the effects on the target area without direct observation of the target. These devices include reflectant focusing devices with diffusing screens and remote closed-circuit television viewing. Recent developments in television screens and tubes permit more detailed close-up and even magnified pictures of the impact area. Closed systems would be especially desirable in laser treatments to the face and for work with high output argon lasers. Such closed systems are used in industrial laser operations.

There are a large number of parameters involved in determining what the energy thresholds would be for causing retinal burns. Color of the retina, wave length of the incident light, degree of beam collimation, whether light is focused in the cornea or not, acceptance area of the eye (dark- or light-oriented), condition of the lens (i.e., how well it brings the laser beam to focus on the retina), energy density of the incident beam, and specific absorption characteristics of the eye other than the retina are but some of the variables involved. A generalized

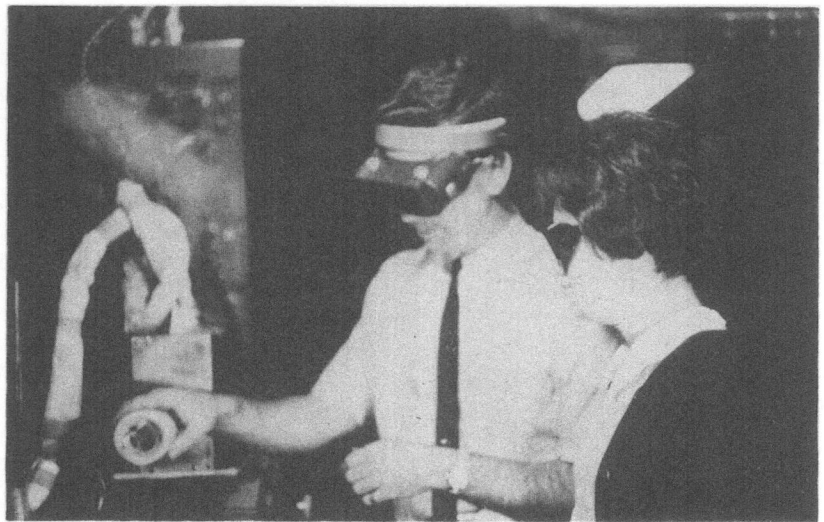

Fig. 20–3. Use of protective glasses during charging of the laser.

formula for determining the threshold energy for retinal damage which incorporates most of the parameters has been developed by Solon. However, it is better to determine under the worst conditions that could occur what quality of filter would be required to give adequate protection to the personnel.

Personnel protection for the eye means a goggle which has sufficient protective material and which is so fitted that stray light cannot come in from any angle. This is especially important for high energy laser work. This means the development essentially of a welder's type of goggle with an efficient filtering glass. This is superior to the spectacle-frame type of goggle where light can leak in from all sides. The protective goggles must be comfortable since some procedures, such as neurosurgery, require use of the goggles for periods as long as four to five hours. We still recommend Schott BG-18 glass with an optical density of 9 at 6943 Å, 10,600 Å and 6328 Å (He-Ne gas laser). Optical density is the common logarithm of the reciprocal of transmittance. A principal ingredient of this is copper sulfate. It is difficult to make a general statement on what constitutes the maximum incident energy density permissible on the glass. However, we use the figure of 100 joules/cm² for our glasses because at that value cracking and cratering become consistently evident. However, the glasses should be designed to give one-shot protection in the event of an accidental direct beam exposure. Swope and Koester have recommended the use of a second filter with a lower absorption coefficient (Schott BG-38 glass) in front of the Schott BG-18 glass to

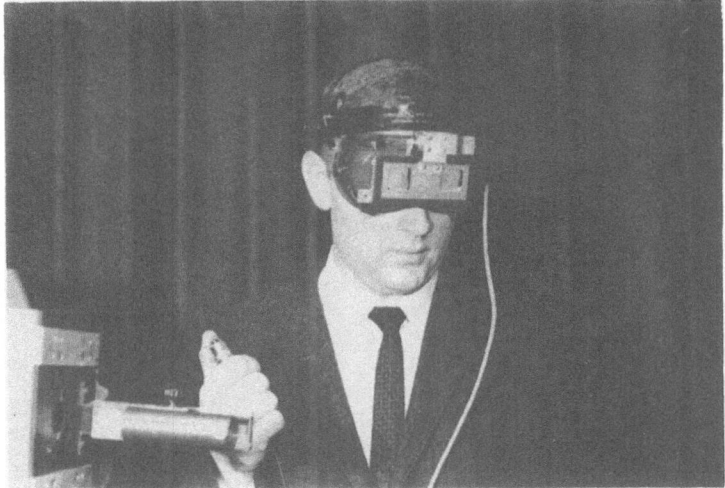

Fig. 20–4. Showing early experimental model of occlusive shutter type of laser fail safe protection glasses connected to laser firing system. Not suitable for CW lasers.

present a higher threshold for crazing and for breakage of the filter. However, this does make the glasses heavier. Despite advertisements to the contrary, there are no single, universal laser eye glass goggles. Glasses should be marked to indicate whether they are protective for ruby, neodymium or argon. Operators of argon lasers must not forget that the argon beams exit from each end of the laser.

These filters will not protect against ultraviolet light lasers or the powerful blue-green light from the argon laser. For these reasons, there is considerable interest and research now in occlusive shutters for protective glasses. There are two types of these—the explosive carbon particles cover and the black shutter. The explosive carbon cover, developed by the Applied Physics Laboratory at Johns Hopkins, was developed to protect aviators' eyes from nuclear flashes. The intense light sets up an explosive charge in the eye glass from which projects a carbon layer over the glasses. The time required (50 to 100 microseconds) is too long for laser techniques where pulses vary from milliseconds to nanoseconds. It has been suggested by Buckingham that the charge be made to cover the glasses before the laser is fired. Another type of shutter arrangement, would be a black shutter to cover the eyes. The solenoid mechanism for the shutter would be connected to the laser firing system. Only after complete closure could the laser be fired. A protype model of these "fail safe" glasses has been developed in our laboratory. The difficulty, as indicated, is with the argon laser and the carbon dioxide laser or other cw lasers where the operating surgeon must watch

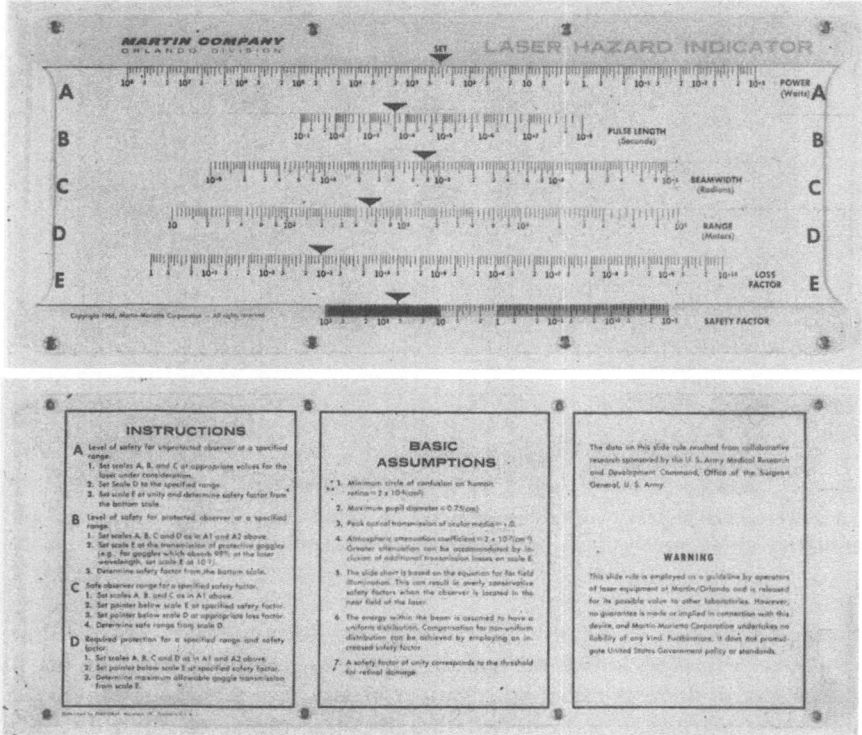

Fig. 20–5A. Laser hazard eye indicator; Martin Marietta Corporation, U.S.A.

the laser operation and still have adequate protection against the intense laser light. In addition to protective glasses, a protective screen of amber colored plastic has been suggested by Bell Laboratories and ruby colored plastic by Flint of Martin-Marietta to protect the surgeon using the argon laser. These preliminary and complex arrangements can be modified as the argon laser becomes more flexible with special probes and a fiber optics system and produces even higher outputs. Many more studies have to be done for protection against not only the argon, but also the carbon dioxide and ultraviolet lasers. For the high output carbon dioxide laser, heavy glass screens and quartz glasses may be used. For this wave length, many surfaces can produce hazardous specular reflection. Skin contact to avoid burn is also important. Even asbestos gloves and aprons have been suggested for the operator. The American Optical Company has developed laser eye protection goggles.

Testing of laser protective glasses may be done by covering a rabbit's eye with the protective glass and then surveying the eyes for damage after laser irradiation. Protection may be observed also with the use of

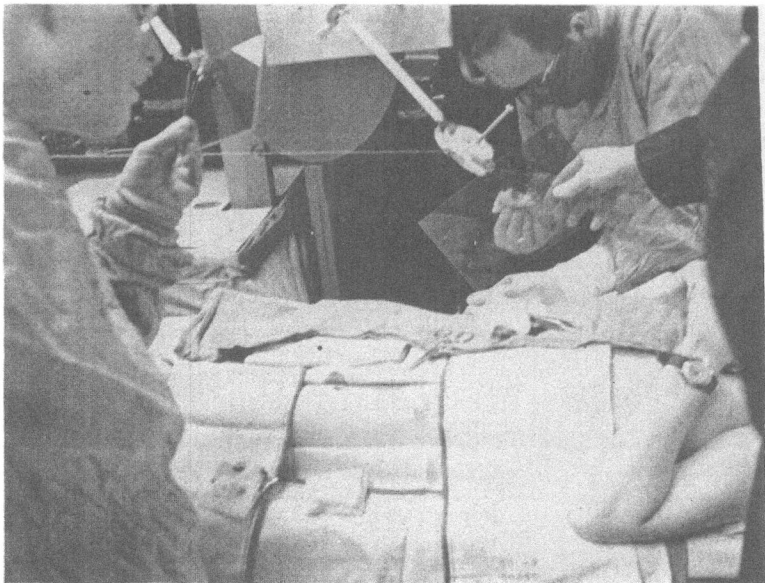

Fig. 20–5B. Use of amber colored plastic sheets for eye protection for surgery with argon laser 2-4 watt output of Bell Telephone Laboratories.

photosensitive detecting devices. Schlickman and Kingston have developed a dosimeter to measure the energy of the reflected laser pulse. This is used in an attempt to prevent eye damage by indicating the light radiation intensity. Calibration is done by experiments on the eyes of rabbits as regards output, pulse characteristics, duration of impact, etc. Finally, eye protection may be done in the least desirable fashion by the evaluation of the operating personnel after use of the protective glasses. Operating personnel should have detailed eye examinations by an ophthalmologist, preferably one who is familiar with laser technology. This eye examination should include an examination and a fundus picture before starting to work with lasers. It is recommended that a fundus camera be available in the laser laboratory to provide a base line picture before laser exposures.

An ophthalmologist familiar with laser technology should be attached to the laser laboratory. As indicated, a detailed pre-employment examination should be done of the eyes of all personnel. This examination should be repeated, including fundus pictures, fundus examination and slit-lamp examination, at intervals according to exposure. Geeraets does not believe it is necessary to repeat fundus photography on routine reexamination unless changes are found which were not present in the original examination. If exposure of the eyes is daily, examinations should be done every three to six months. In case of accidents, an examination should be done at that time. A committee on laser eye protection established by the National Research Council, according to Geeraets, will soon give data on the specific requirements of such examinations, definitions of eye damage from laser, data on so-called "safe threshold" values and requirements for eye protection. We trust these will consider the almost unknown factors of chronic exposure.

In the meantime, there are no figures available as to figures for eye exposure which may be "safe." Jones has suggested 10^{-6} watts per square centimeter for gas lasers and 10^{-8} joules per square centimeter for pulsed lasers. Kaufman's figures are different (Table 1).

Table 1

LIMITS OF SAFE EXPOSURE TO LASER LIGHT (KAUFMAN)
Maximum Safe Levels

Laser Operation	Power Density Incident on Retina	Equivalent Input to the Eye
CW ($t_p > 0.1$ sec)	0.35 W/sq cm	2.7×10^{-7} W
Normal Pulsed $\approx 500 \ \mu$sec	3.2×10^{-2} J/sq cm	2.5×10^{-8} J
Q-switched Pulsed $\approx 300 \ \eta$sec	2.7×10^{-3} J/sq cm	2.1×10^{-9} J

This table is based in part on measurements with ruby and Nd lasers. For extrapolation to other wavelengths, the data on spectral transmission of ocular media of the human eye should be included since tissue absorptance influences the degree of photocoagulation and consequent damage. These threshold values should be considered accurate only to within an order-of-magnitude tolerance. Greater accuracy is precluded by case-to-case variations in the following key parameters.

1. TISSUE

Pigmentation: varies with the individual and location within the eye.

Special absorption: maximal for wavelengths of about 400 to 550 mμ; diminishes toward the infrared; not much known about the ultraviolet.

Blood circulation: Available blood-flow acts as stabilizing mechanism tending to maintain tissue temperature.

2. EYE

Pupil (iris) size: variable aperture controlling the fraction of incident light to be admitted (assuming its diameter is less than that of the incident beam); may also affect retinal image size.

Spectral transmission: very high from 400 to 900 mμ and high at several infrared bands.

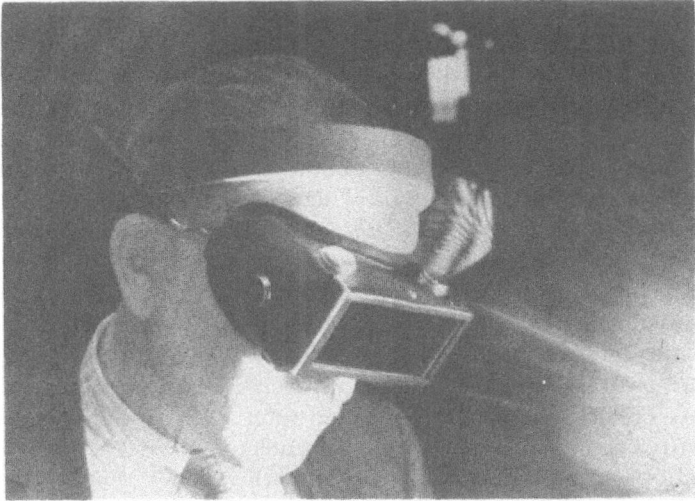

Fig. 20–6. Showing exposure of face on reflectance of laser beam from target area, temple area must fit tightly. Black-and-white copy of picture of laser impact color infrared Ektachrome (*Eastman*).

Covergence power of the cornea and lens: affects the location of the focal plane and retinal image size.

3. ENVIRONMENT

Laser parameters: wave length energy, power, pulse duration and rate, wavelength, beam divergence, transmission devices used.

Location: eye in near or far field of source's effective aperture; interposition of lenses between laser and eye; scattering properties of targets.

Spatial homogeneity: imperfections in laser crystals; multi-molding; atmospheric anomalies.

For their laser operations, Martin-Marietta Corporation has published an excellent safety guide on laser protection. Included in this is an Eye Hazard Nomogram to estimate the safe range "at which a subject may be illuminated directly." The divisions include a scale for loss factors, losses in the eye optical surfaces external to the laser, exit mirror and atmospheric losses, a division for pupillary size, for exit energy of the laser for energy density in the retina and beam width in milliradians. Finally, the safe range is given.

The development of ultrasonic (elastic recoil) waves, especially in cavities such as the eyes, must be considered also. Such waves have been detected by Amar on the back of the skull after laser impact of animal eyes. Many experiments are under way now with transducers to detect such waves in all areas of the body.

Infrared color photography with special film from the Eastman Kodak Company (Ektachrome Aero, Type 8443) has been used by us to study reflectance from target areas with laser impact both from ruby and neodymium lasers as well as argon and carbon dioxide. This type of photography done in a dark room using a wide-angle lens, can show the detailed pattern of reflectance on personnel, including protective glasses and exposed skin.

Many rumors have been heard about types of corneal opacities or cataracts with high energy (especially neodymium) and high peak power Q-switching. As reported previously, Jones had produced severe eye damage in primates using high energy laser impacts.

For biomedical purposes in direct impacts about the face of patients, we have used extra black cloth coverings for the eyes in addition to the protective goggles. A double felt curtain withstands up to 100 joules/ cm^2 from a direct impact without appreciable transmission. It is the experience of the patients who have had such impacts that some reddish light is still perceived as it is with the protective glasses alone (possibly by refraction of the light as it passes into the tissue). For neodymium, Buckley has shown considerable absorption of the neodymium light in this black material. The tapered curved quartz rod for the transmission of the laser beam may reduce some absorption and reflectance away from the target area. It must not be forgotten that eye protection must be considered in all work with holography.

Under development in our laboratory is a laser protective shield which can be inserted directly over the eyeball under local anesthesia.

It is recommended that as a rule personnel not work in the dark since the difference in eye acceptance area between being in the dark and in the light is 16:1.

As indicated, related to eye protection is the transmission of light to the eye or orbit by soft tissue, the transillumination effect with lasers. This is important in accidents which may occur from impacts about the face, such as with premature firing of the flash tubes, etc., and in treatment impacts of personnel about the head and neck of a patient, including laser dental surgery. Laboratory personnel should be careful in working with high-output helium-neon gas lasers for transillumination. We have attempted to study this in patients by measures to avoid surface reflectance and, as mentioned above, with detailed photography especially with the new Kodak colored infrared film. Black cloths and various materials are used to attempt to protect the eyes and the face from the reflected beam and the photographs are taken of the impact in darkness. The photographs of the protective areas are taken before, during and after impacts with either Kodachrome, Ektachrome or the new experimental Kodak colored infrared film. Exposure conditions for this film have been determined by Buckley and described in the chapter on Laser Photography. In brief, in exposures about the forehead and the buccal cavity for laser dentistry, it appears that there is transillumina-

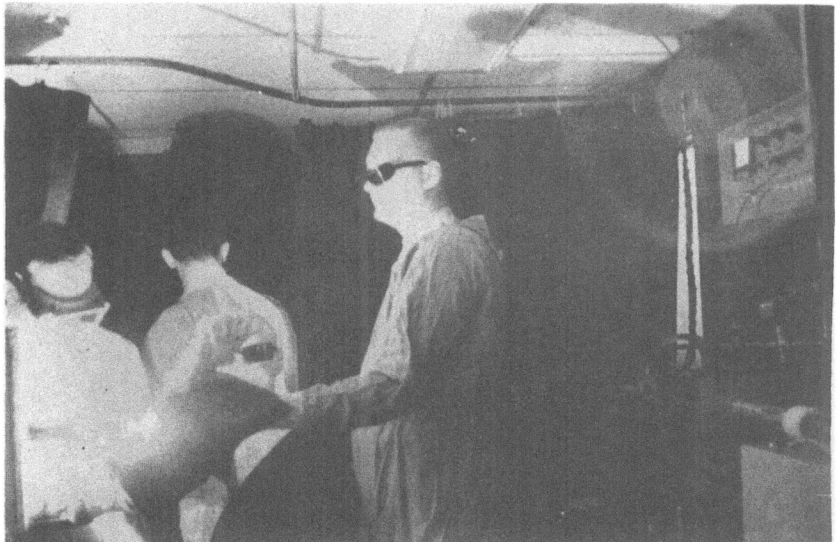

Fig. 20-7. Ruby laser impact on squamous cancer of forearm; black-and-white copy of color infrared picture with wide angle camera lens, showing extent of reflectance throughout entire room.

tion of the sinuses and also of the soft tissues. As yet, no permanent sequelae can be recognized. As mentioned previously, this is an important study and should be done before the biomedical applications of the laser become significantly more widespread and especially before the laser is used for actual transillumination procedures.

In summary, then, as mentioned previously, an eye protection program should be constantly maintained and any program must be continually re-evaluated in the light of continuing studies on the immediate and delayed reactions of the eye to laser radiation. As described in the chapter on Eye, there have been serious accidents from laser impacts. Constant vigilance must be maintained even during charging procedures since premature firing may occur. It should not be forgotten also that the xenon light itself, because of its intensity, can cause eye damage.

SKIN PROTECTION

Another concern in personnel protection with the high energy and high output lasers is the exposure of the skin. The direct impact of high energy lasers may cause considerable damage to the skin, especially where it is pigmented. Recently, studies have been done to measure this reflectance by a rapid scanning spectrophotometer. The skin has been checked in visible light and with examination under Wood's filter to de-

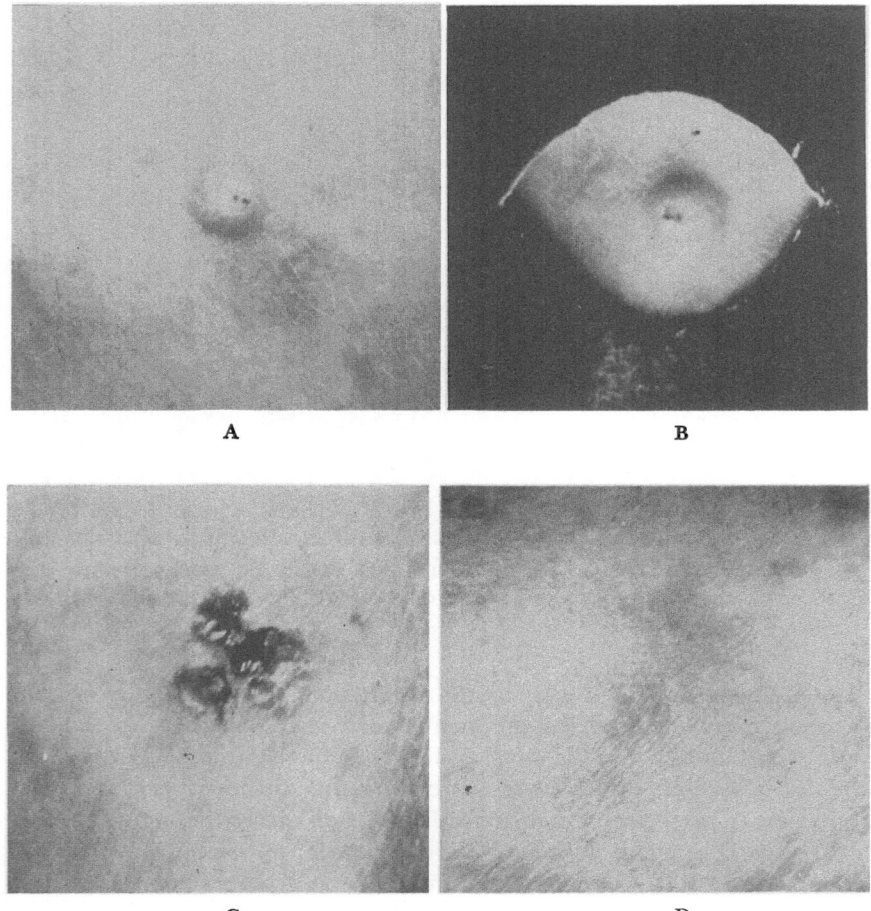

Fig. 20–8A. Squamous cancer of skin on forearm. **B.** Showing use of protective black cardboard. **C.** After laser impacts, pulsed ruby laser, exit energy density 1,000 joules/cm², showing protection of the skin with cardboard. **D.** Showing reaction in two weeks. Later this cleared, and biopsy was negative.

tect any early pigmentation or dryness or scaling. There is, as yet, little data known of repeated exposures of exposed areas from reflectance. In our laboratory, with exposure periods varying up to more than four years, no changes from chronic exposure to the skin have yet been found. In one research worker, an individual with a hyperreactive atopic skin, deliberate attempts were made to sensitize this individual. At present, even low energy densities in the nature of 0.23 joules of exit energy and 18 joules/cm² energy density in a target area of 0.004 cm² produce significant reactions in the skin. At present, we are concerned with skin exposure during operations with high output argon lasers. Surgeons' hands, unless protected by gloves and sleeves of the operating gown are exposed for prolonged periods to the laser. The face outside of the protective glasses area is also exposed unduly. Where patients are receiving impacts with the laser, the personnel in the laboratory who will be called on to assist in positioning the patient and will be exposed to numerous impacts, should attempt to keep their hands and face protected. Black leather or black felt gloves or felt coverings can be used on the hands. The face should be turned away from the target area. As indicated above, pictures of the impact in a dark room with infrared colored film reveal the extent of areas of exposure about the target zone. Only with long-term studies will it be possible to determine what the chronic radiation effects are on the exposed skin. Liquid crystals, because of their color-changing property, and photochromic dyes are under investigation for the skin protection program. Experiments are under way to develop laser irradiation monitoring badges with liquid crystals in plastic. Patterns of color change about the target area may reveal the extent of even minimal laser effects.

In high energy laser treatments of skin malignancies, the skin around the target area has been protected by cardboard used directly on the skin and over aluminum foil. Cardboard is found to be effective alone. Experiments are under way with the colored liquid protectant films to be applied to the skin for protection of irregular areas.

The increased absorption of the laser beam, including that of the argon laser, in blood means also that direct impacts should be avoided on the superficial blood vessels in the skin of laboratory personnel. Detailed hematologic examinations in our laboratories have failed to reveal any blood disturbances in laser personnel over a period of two years.

AIR CONTAMINATION

Air contamination is also a problem especially as it relates to scatterings of plume fragments, Raman and Brillouin scattering. Previously, nitrogen concentration was of concern. The concentration of nitrogen vapor could become significant in confined spaces with a consequent reduction of oxygen concentration. Preliminary measurements in our laboratory by the Occupational Health Field Headquarters of the Public Health Service have shown no low oxygen levels. The increasing use of water-cooled lasers in preference to the nitrogen-cooled types does reduce the frequency of the use of nitrogen. Liquid nitrogen also produces burns when handled in a careless manner.

In our experiments with peak power outputs of 100 to 150 megawatts, the search for x-ray and grenz ray generation in various targets, both

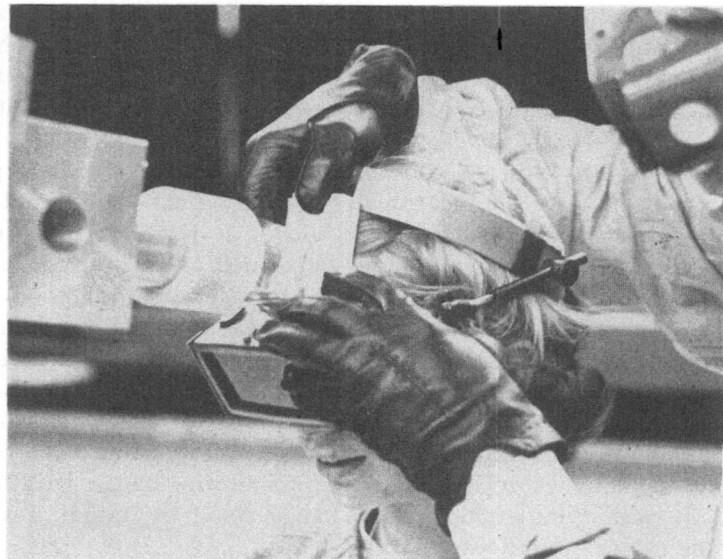

Fig. 20–9A. Showing protection of operator's hands by using gloves and protection of forehead with cardboard template and quartz glass plume trap for treatment of linear portwine angioma of forehead in 12-year-old-girl.

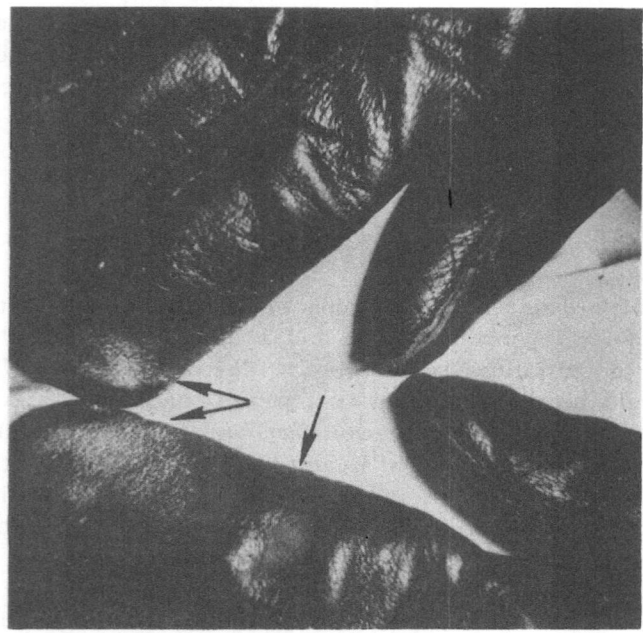

Fig. 20–9B. Showing bleaching of black protective gloves held near target area.

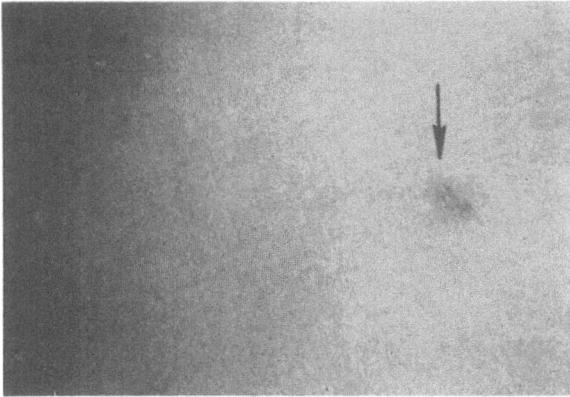

Fig. 20–10. Deposit of silver on skin after impact; pulsed ruby laser, 18 joules/cm² energy density, unfocussed beam, of black film strip used on skin as protectant.

metallic and non-metallic, by the radiation physicist, Keriakes, failed to elicit any evidence of this, although such generation has been reported to us. Our methods of analysis were crude for this type of technique, i.e., essentially a Geiger counter and the use of x-ray sensitive film. Recently, in much more significant experiments, Schwartz has reported no radiation traces from peak power outputs of 20 to 30 megawatts in a cloud chamber. However, this report did not analyze the interaction with metals. We have used high peak power outputs directly into cloud chambers to attempt to detect specific tracts. As yet, none have been found. Biological testing, such as cytogenetic studies and electron spin resonance spectrometry, will have to be done. We have done preliminary controlled experiments with electron spin resonance spectrometry after laser impacts on living skin of man with controlled experiments with ultraviolet, x-ray, Grenz ray and electrocoagulation. Jacobi has detected signals for us only in pigmented tissue. These studies are continuing.

For some years, we have been concerned with air contamination by plume fragments. These particles of viable tissues are projected from the target area. We have studied their distribution pattern with infrared color photography, fluorochromes in the target area, tissue culture and bacterial culture techniques. The simplest plume trap we have used is the quartz glass cone about the laser head and target area. Riggle and Hoye have developed a vacuum cylinder and a polyethylene bag about the tumor area. Plume traps are necessary primarily for personnel protection.

As indicated in Chapter 7, our laboratory is conducting experiments on the air pollution and fire hazards of laser Raman spectroscopy with such materials as benzene, toluene and carbon disulphide. Proper venti-

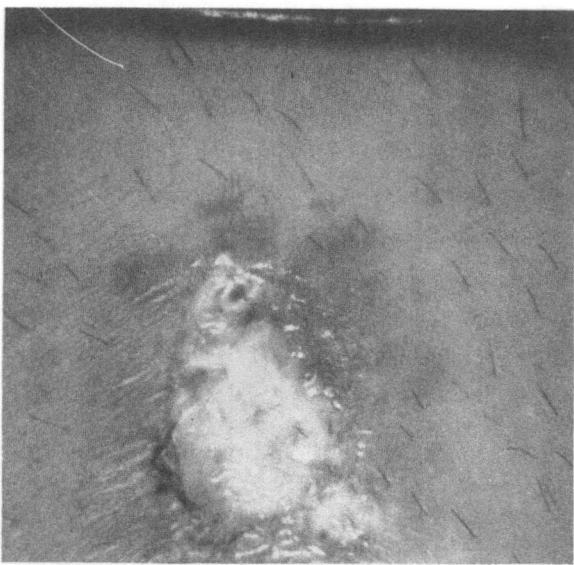

Fig. 20–11. Electrical burns on thigh of personnel from exposed wire in laser laboratory.

lation, fire safety measures, and if possible, the selection of samples with higher boiling points are necessary. Kronoveter has studied airborne concentrations of benzene and toluene in our laboratory after Raman spectroscopy. At exit energies of 50 joules, concentrations of benzene varied from 1.1 to 1.4 ppm and toluene 2.9 to 11.4 ppm. For high output systems, exhaust systems may be necessary.

Ozone is produced at times about the flash lamps and concentrations of this could build up with high repetition rate lasers.

Because of the extraordinary degree to which the laser beam can be collimated, this light will always be a hazard. Damage threshold relates to many factors such as energy and power densities, pulse length, pulse characteristics, transmission of laser beam, the characteristics of the target areas whose changes have to be measured both early and late. Because the precise limit of these changes is still difficult to measure, it is difficult, at present, to define accurate damage threshold levels. Only approximate ranges can be given.

The chief hazard continues to be the eye. Skin protection is also necessary. Air contamination then may result from such conditions as decreased oxygen concentration with liquid nitrogen, contamination by plume fragments from the target area, chemical and fire hazards from laser Raman spectroscopy, ozone and possible ionization of air. Electrical shocks may occur also. Medical examinations of laser operating personnel should not be limited to the eyes, but should include the exposed

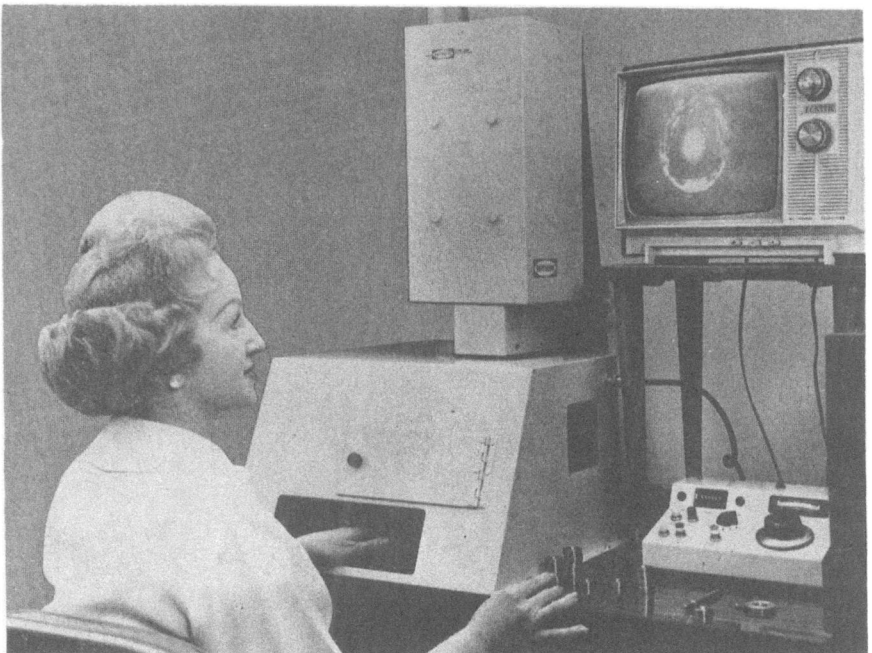

Fig. 20–12. Showing protection of operator in production-diamond piercing by laser through remote circuit television monitoring. (*R. W. Cawley-Western Electric Co.*)

skin, the chest and blood counts. As yet, there is no data for man indicating that the initial changes in tissues observed from laser research of the past four years are similar to those produced by x-ray. However, the late effects of the laser reactions are not known. Therefore, adequate records and follow-up examinations will be required. This is especially important for industrial medicine where toxicologic studies of plume fragments from metal impacts are just beginning.

Chest examinations with x-ray studies have shown no changes in our laboratory personnel.

ELECTRIC SHOCK

In the research laboratory, the possibilities of electrical shock are great when space is at a minimum. Some high energy systems may require upwards of 50 capacitators. The energy stored at high voltage in these capacitors when dissipated through a human conductor will cause severe shock and massive thermal burns. This caution applies also to the cavities of some lasers which are maintained at the high voltage side of the system.

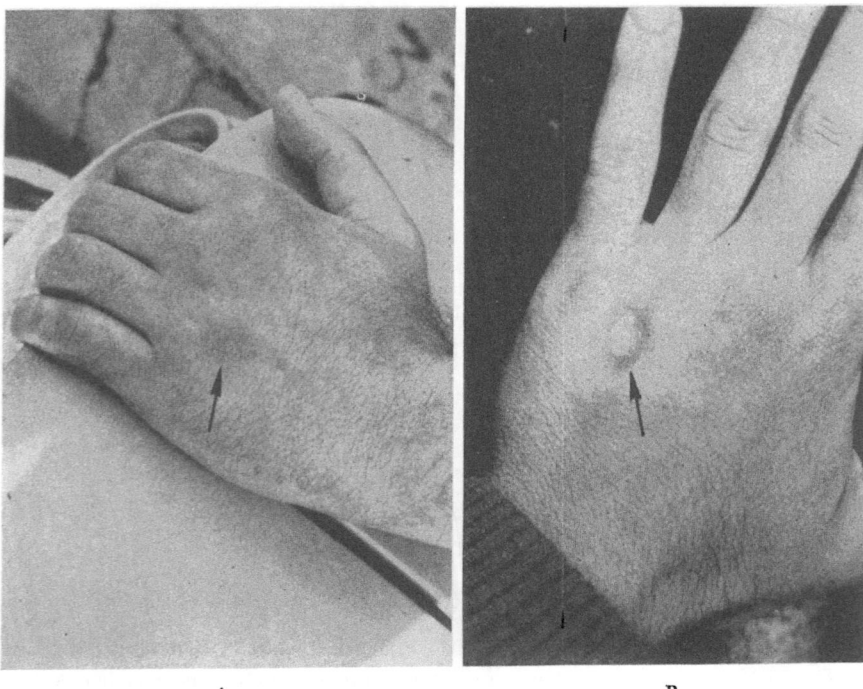

A   B

Fig. 20–13. Accidental exposure dorsum of hand; 500 megawatts peak power output Q-switched ruby laser. A. shortly after exposure; B. after two weeks. (*Patient and photographs courtesy of Dr. Louis Sperling.*)

FOLLOW-UP STUDIES

With the late effects of the laser on living tissue as yet unknown, and with the increasing development and use of high energy and high power output laser systems, it is necessary to set up long term study programs both on patients who have been treated and on laser personnel. Impacts must be described, as listed previously, in great detail. In our patients treated with laser radiation in the last four years, there has been no evidence of complications other than nonspecific scarring. These have been studied in some experiments with high output lasers. The increased reactivity of one research worker to low energy laser radiation has been mentioned above. Personnel records should include estimates of daily exposure, reports of eye examinations, fundus pictures and other laboratory data. Any accidents should be recorded in detail. As advocated by others, a central registry file of laser accidents should be kept.

As in any industrial process, personnel unfamiliar with the details of the process can get into the most difficulties. Those who use the high energy and high power output lasers should receive detailed instruction in actual use and not rely on an instruction manual. Instruction manuals for laser technology, in our experience, are often incomplete in details of protection and often in details of construction and assembly.

CONCLUSIONS

A planned safety program and laser safety officer are necessary in any laser installation, no matter how small. With the development of new and higher output lasers, the laser safety program constantly must be reviewed and revised. With such programs, meaningful investigative studies with the laser can be done in man.

REFERENCES

Amar, L.: Personal communication.

Babla, J., and J. John: The saturation effect in retina measured by means of a helium-neon laser. *Amer. J. Ophthal.*, **62**:659, 1966.

Buckingham, B. H.: Explosive light filter. Applied Physics Laboratory, Johns Hopkins University. Personal communication.

Cullom, J. H. and R. W. Waynant: Determination of laser damage threshold for various glasses. *Appl. Opt.*, **3**:989, 1964.

Daniels, R. C. and B. Goldstein: Lasers and masers—health hazards and their control. *Fed. Proc.*, **24**:27, 1965.

DeMent, J.: Personal communication.

Fine, S., E. Klein, Glen Hardway, R. E. Scott, W. King, and C. Aaronson: The use of colored circuit television in laser investigations. *J. Invest Dermat.*, **42**:289, 1964.

———— W. Nowak, W. Hansen, K. Hergenrother, R. E. Scott, J. Donoghue, and E. Klein: Measurement and hazards on interaction of laser radiation and biological system. *Nerem Record*, **6**:158, 1964.

Geeraets, W. J.: Personal communication.

Gibson, H. L., W. Buckley, and K. E. Whitmore: New vistas in infrared photography for biological surveys. *Jour. Bio. Photographic Assoc.*, **33**:1–33, 1965.

Goldman, L.: Protection of personnel operating lasers. *Amer. Jour. of Med. Electronics*, **2**:335–338, Oct.–Dec., 1963.

———— and P. Hornby: The design of a medical laser laboratory. *Arch. Environ. Health*, **10**:493–497, March, 1965.

———— ————: Personal protection from high energy lasers. *Amer. Indus. Hygiene Assoc. Jour.*, **26**:553, 1965.

———— and D. Richfield: The effect of repeated exposure to laser beams: Case report with nine months' period of observation. *Acta Derma.-Venerol.*, **44**:264–268, 1964.

Harper, D. W.: Laser damage in glasses. *Brit. J. Appl. Phys.*, **16**:751, 1965.

Jacobi, F.: Personal communication.

Jones, A.: Personal communication.

Kaufman, Jesse C.: Laser technology. April, 1966.

Kronoveter, K. J.: Personal communication.

Martin-Marietta Corporation. What you know about laser safety. 1965.

Riggle, G. C. and Hoye, R. C.: Some engineering problems in the use of a high energy neodymium laser on malignant tumors. *Engineering in Medicine and Biology, Proc., 19th Annual Conf.,* 1966, p. 58.

Schlickman, J. J. and R. H. Kingston: The dark side of the laser. *Electronics,* April 19, 1965.

Schwartz, J.: Personal communication.

Solon, L. R.: Occupational safety with laser (optical maser) beams. *Arch. Environ. Health,* 6:414, March, 1963.

Sperling, H. J.: Personal communication.

Straub, H. W.: Protection of the human eye from laser radiation. Report from Harry Diamond Laboratories, Army Materiel Command, July 10, 1963.

————: Protection of human eye from laser radiation. *Ann. N.Y. Acad. Sci.,* 122:773, 1965.

————: Use of protective goggles in areas of laser radiation. *Fed. Proc.,* 24:78, 1965.

Swope, C. H. and C. J. Koester: Eye protection against lasers. *Applied Optics,* 4:523–526, 1965.

Terbrock, H. E., W. N. Young, and W. Machle: Laser medical industrial hygiene controls. *J. Occup. Med.,* 5:564, 1963.

Zaret, M. M.: The laser hazard. *Arch. Environ. Health,* 10:629, 1965.

21

Design of the Biomedical Laser Laboratory and the Laser Operating Room

Laser research requires separate and planned facilities. This is true at least for biomedical applications. Furthermore, at least for the present, investigations in effects of the laser on living tissue will have to be controlled as much as possible. Such controlled research can be done and adequate precautions taken in areas devoted solely to laser work. If only an optical bench is used somewhere just in the corner of a room, it is difficult not only to work in this restricted area, but also to provide the special safety measures which laser research requires.

The treatment of patients complicates this set-up. Here the laser is a complex addition to the treatment center, and the detailed operatings of the treatment center and the operaions of the laser must be combined.

THE BASIC REQUIREMENTS OF A BIOMEDICAL LASER LABORATORY

Even if the laser research is conducted on a modest basis, and even if the laser center is, as usual, restricted to a limited space, an effort should be made still to have even a separate small room for laser experiments. This should be a well-ventilated room painted black, with dull reflectant surfaces, provided with doors which can be kept locked during laser experimentation and marked by an appropriate "Laser-Danger" sign. If laser research is limited to laser microscopy then the working unit can be very simple. If biomedical laser studies limit the laser solely to an optical bench, the set-up continues to be simple. If laser is to be

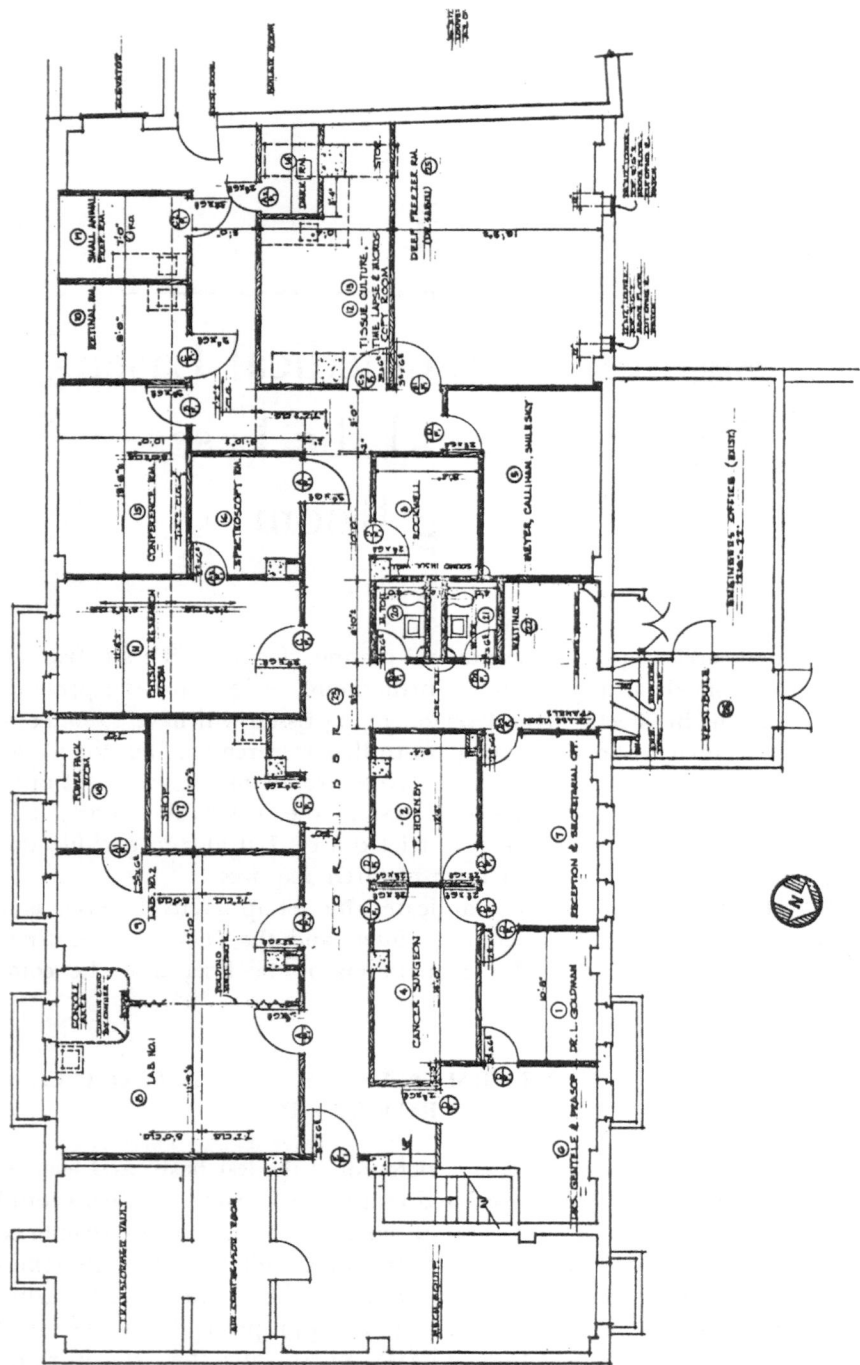

Fig. 21-1. Design of Medical Laser Laboratory, Children's Hospital Research Foundation, Cincinnati, Ohio. Supported by the John A. Hartford Foundation.

flexible with high outputs, then larger facilities will be necessary. Someone should be appointed to be in charge of the laser safety program.

LASER INSTRUMENTATION

If a complete biomedical laser unit is to be established, then it must combine the following basic features: (1) biomedical research; (2) laser physics research; (3) adequate safety measures; (4) provisions for expansion.

If the research is to be limited to embryology, cytology, and cytogenesis, then the laser accessory to a microscope is the basic instrumentation. If the work is limited to one wavelength, then the laser attachment to the microscope as indicated previously would be relatively simple. If different wavelengths are to be used, then a whole series of laser installations may be necessary. The different lasers may be direct parts of the microscope or used as external sources and reflected through the microscope optics. Safety precautions are of course more complex with the latter type of arrangement.

If the microscope research demands impact areas of the magnitude of 1 to 2 microns in diameter, then the instrumentation will be more sophisticated, especially if the laser energy densities of 50 joules/cm^2 or more are desired.

If the biomedical laser research group is just initiating laser research, it is well to begin with low energy laser instrumentation on an optical bench. If the program plans to include studies with high outputs and high peak power outputs, then the whole installation has to be planned and developed.

It is obvious that such installations will be very expensive, both as regards instrumentation and construction. Basic area for a detailed program on the biomedical applications of the laser requires at least 2500 square feet with facilities for a staff of at least 20. There must be adequate electrical supply currents available, adequate ventilation and even adequate planning for the future.

The essential features of this laser research area may be listed briefly as:

1. Laser operating areas
   a. Ambulatory patients
   b. Eye patients (special facilities)
   c. Animal surgery
2. Laser microscopy
   a. Laser microscope and attachments including microscopic holography.
   b. Histopathology
   c. Tissue culture

3. Laser chemistry and biochemistry
4. Physics research
5. Instrumentation, repair and maintenance

With all these facilities, safety measures must be provided.

The laser treatment room for patients should contain laser flexible equipment so that patients may be treated either in a treatment chair or on a treatment table. This means equipment flexible enough so that any area of the body may be treated, including the mouth for laser dentistry. The laser head may be attached to special stands, trolley systems or to frames. For the present, at least, treatment rooms must contain facilities for the essential laser instrumentation for biomedical research: the ruby, neodymium, and argon lasers.

Laser eye treatment should have a separate room with the retinal laser instrumentation fundus camera and treatment table and other facilities required by the laser ophthalmologist.

The animal treatment room is the room where the animal is prepared for laser surgery and then moved to the laser treatment room. If the laboratory can afford it, sometimes special laser equipment such as a newly developed experimental pilot system may be able to be set-up initially in the animal operating room.

Requirements for the laser microscopy room are obvious. As indicated, difficulties arise when it is desired to use different wavelengths, energy densities and power outputs. If facilities for laser spectroscopy are available these should be incorporated in this unit or in the laser chemistry unit. In all laser microscopy units, due attention should be given to damage of the optics of the microscope, especially as power and energy outputs are increased. Time-lapse photomicrography units are important accessories in this installation, especially in work with tissue cultures. Phase contrast microscopy is preferred for tissue culture studies.

Laser chemistry and biochemistry facilities should be related to the biomedical laser laboratory program. Some parameters of research in this area would include work with dyes, studies on DNA, RNA and amino acids, preparation and standardization of tissue extracts for cancer work. In our laboratory, cancer immunologic research is combined with this laser facility.

Physics research is an important part of the biomedical laser installation. One example is research on measuring outputs. The development of new instrumentation constantly in laser research should be studied first in the physics laboratory. This is also the place where laser physicists have an opportunity for continued basic research. Otherwise, the physicist becomes just a supervising technician and maintenance man for the laser laboratory installation.

LASER OPERATING ROOM

An important recent development of the laser is the development of the laser operating room. Previous chapters show the rapidly expanding field of laser surgery. Briefly, this now adds the complexities of the already complex environment of a major operating room.

Laser installations may be simply additive to the operating room, or the entire operating room may be designed as a special laser operating room. For most installations, the laser operating room will continue to be an operating room to which the laser is brought or is used as a type of permanent installation. As research in laser surgery continues, then specific laser operating room facilities will have to be developed. One example of this separate facility is the laser operating room of Ketcham now under construction at the National Cancer Institute in Washington, D.C.

As indicated in previous chapters, special laser instrumentation must be built into the operation room. This is necessary at least for attachment of the laser head. In our operating room, a ceiling-mount laser head (Applied Lasers Division of Spacerays Inc.) is under direct control of the surgeon.

As shown in previous chapters, lasers in the operating room will be, for the most part, lasers of high energy and high peak power outputs. They will demand safety precautions on the part of the entire operating room personnel. This complicates the operating room equipment since surgeons and nurses will have to wear protective glasses. The flexible operating room laser unit must be protected as regards plume traps, accurate focussing and fixation of the laser head, uniform control of output, and conformity to selected requirements of operating room instrumentation. The hazard of electrical shock is not as great as one would imagine since instrumentation is mostly confined except at the target area. Here the same precautions hold as for the laser rival, electrosurgery.

The power supply is kept usually out of the operating room and cables are attached to the flexible laser unit. Provisions should be made for equipment replacement so that the laser surgery can be continued if glass tube or rods fail. It has been indicated too often that the laser is not assembly line equipment, and its performance may be erratic. So, replacements should be available readily and rapidly within the laser operating room. In our experience, planned programs with the physicists, technicians, and the surgeons develop a team for operating room laser surgery.

CONCLUSIONS

It is obvious, then, that laser facilities for biomedical research must be planned, not done blindly. The goals of one's program must be known so that this facility can be planned to follow this program. Laser facilities should be separate units especially in the use of high output equipment. Safety measures must also be planned in detail in any area where lasers are used. Since laser research is still in its infancy, planning for expansion should always be considered.

REFERENCES

Goldman, L., P. Hornby, and J. Solsman: A design for a complete medical laser laboratory. 16th Annual Conference on Engineering in Medicine and Biology, November, 1963.
———, ———: The design of a medical laser laboratory. *Arch. Environ. Health,* **10:**493, 1965.

22

Present Status and Future Developments of the Biomedical Applications of the Laser

It is well at this time to consider where we are, and then where we plan to go with biomedical research. It is obvious that the laser still has a dramatic appeal for the public even though, as yet, it has not fulfilled many of the early reckless dreams of fiction writers. It is obvious now also that lasers, especially those with the high exit energy and high power output, are very expensive. Moreover, laser instrumentation has still not reached the desired state of maturity for which the early investigators prayed. So, biomedical research has not kept pace with the other fields of laser research.

LASER IN SCIENCE AND INDUSTRY

There is no doubt at all about the place of the laser in the modern physics laboratory. There the laser was developed and there it will always be. Here, the basic advances in laser physics will continue to be made.

Many nations are, and will continue to be, interested in laser weapon technology and in military ranging, missile tracking, communications, etc. Industry will continue to use the laser for precise measurements, alignments of optical instrumentation, communications, welding, drilling of refractory materials, wheel balancing, construction work, and holography.

BIOMEDICAL APPLICATIONS

There have been encouraging results in the field of biology and medicine. To us, it appears that laser microscopes are a worthwhile accessory in microscopy. This instrument provides opportunities both for basic studies in photobiology at cellular level and for a marvelous precise microsurgical instrument for cellular surgery. Photobiologic studies can be done with lasers of various wave lengths, varying energy and power outputs, and diverse transmission systems. Laser microscopic holography is now available. A whole new field has been opened up in experimental embryology, cytology and cytogenetics. The microscope laser is also an excellent accessory in the field of spectroscopy. This instrument expands the direct use of spectroscopy for living tissue and for accurate area analysis. Whether quantitative data is as precise as with other methods of spectroscopy is still questionable, but this will be resolved in the future. The recent change to a neodymium laser in the laser microprobe will help.

A brief note may be said about the surgical technology today. With routine surgical techniques used in most areas of surgery, a plateau has been reached in technical achievement. The surgeon now turns to the engineer for help. At present, there are three types of instrumentation that should be included in a surgical research program: (1) cryo-surgery—liquid nitrogen probes as surgical tools; (2) laser surgery and (3) the plasma torch. With the rapid developments of the argon lasers, precise cutting lasers are being developed. This can be done if efficient flexible probes and fiber optics systems can be devised to transmit high power outputs of the argon and other gas lasers.

In ophthalmology, the laser has limited value for the treatment of retinal tears, detachments, angiomas, melanomas and retinoblastomas. For the laser retinal coagulator, the xenon photocoagulator is still its great rival. Only continued research will establish which is superior for any particular area. Current interest is in attempts to prevent vascular damage, as in diabetes, by thrombosing blood vessels by the laser in the treatment of uveitis, chorioretinitis, and edema of the macula.

In dermatology, laser surgery offers facilities for basic research in new parameters of radiobiology of the skin and for the rapid treatment of multiple malignancies, for the treatment of melanomas, for continued investigative studies in the therapy of angiomas, especially portwine types, and for treatment of warts and various pigmentations of the skin. Laser surgery of tattoos should be continued. The ease of the argon laser and Q-switched laser cutting their way among the tattoo designs is of great interest and value.

In general surgery, the laser is used for the treatment of melanomas, for investigative studies in vascular surgery attachment of grafts and for metastatic tumors. Laser research in the surgical specialties of urology, orthopedics and in dentistry is just beginning.

Additional investigations continue in cancer therapy programs with the hazards of the laser in spreading cancer, the need of adequate controls, the studies on the synergistic effects of laser and cancer chemothera-peutic agents and the laser and other modalities of radiation.

Throughout this book it has been emphasized repeatedly, that since the laser is still an investigative tool, controls must be used wherever and whenever possible. At present, these controls include electrosurgery, excisional surgery and also other modalities of radiation, especially xenon and the plasma torch.

FUTURE DEVELOPMENTS

The basic patterns for future developments of the biomedical appli-cations of laser technology can be listed briefly as: (1) the development of laser instrumentation including laser holography; (2) the continued research on basic mechanism at cellular levels and (3) continued con-trolled clinical studies at special laser medical centers; (4) the develop-ment of an industrial hygiene program for lasers in industry; (5) as the result of all these studies, the development of laser technology discipline in biomedical engineering.

Laser instrumentation must be more flexible so that it can be used freely as a knife whether it be on the microscope or in the hands of the surgeon. An example of this, is the use of special flexible probes with the argon laser. There should be a wide range of wave lengths available for the particular problems and the instrumentation should be reliable and constant in output, even with repetitive firings. There should be adequate, precise and standard means for measuring outputs, both low and high. Finally, the development of instrumentation should be such that the costs not be prohibitive and not interfere with con-tinued expanded research and development.

Although adequate examples have been given for the existence of factors other than thermal in the laser reaction, too many believe from the casual acquaintance with the laser that there is nothing to this instrument but thermal outputs. Continued research must be done then to define not only the parameters of the thermal component, but how to measure it accurately and rapidly.

There must also be definitions of the parameters and true significance of the ultrasonic and pressure shock waves on living tissue. What is their order of magnitude? What does cavitation do to living tissue mass, etc.? The question of free radical formation must be established by better controlled experiments and by the use of more sensitive equip-ment. Ionizing radiation after laser impact must be sought for in the air and in tissues. Chemical changes are of significance, also; for example, in the brilliant fluorescence of the living tissue of the argon laser. The value of the techniques of tissue culture experiments and animal ex-periments for basic laser research have been shown repeatedly. All these should be continued so that the concepts of wave length dependencies and energy density requirements for the cell, for tissue masses and for specific cytogenetic changes can be determined. Studies in laser radiation should be as complete, at least, as those done by other forms of radiation. Holography for medicine and biology is just developing: cellular kinetics, sonar holography are but a few phases.

Laser chemistry, laser microbiology, and laser botany are still very much in their infancy. For fundamental studies in photobiology, botany offers much. Laser microbiology provides a technique for basic studies in enzyme disturbances and cytogenetics. Laser instrumentation will be applied even more in studies in all fields of photochemistry including now plasma chemistry.

It is emphasized again that biomedical research requires the establishment of specially designed biomedical laser laboratories in medical centers. Such laboratories can develop laser treatment centers through the efforts of continued research, development, servicing, etc. Laser treatment centers should be developed as separate and distinct facilities in a medical center so that they can expand as laser developments occur.

Continued research must be done on all phases of safety precautions for the laser radiation. This should include such features as skin, blood vessels, eye, and air pollution studies.

All this means, then, that expanded basic studies with lasers must continue. For progress, this requires adequate interest and actual participation by all the agencies concerned and still the expenditure of considerable funds.

23

Summary and Conclusions

As one attempts to review the data on the biomedical applications of the laser, one finds much of it, especially as applied to medicine, to be preliminary probing and consequently, incomplete. At least at the present time, a wide range of laser instrumentation and wide range of outputs are being investigated. These include the following: (1) the ruby; (2) the neodymium; (3) the argon; (4) the krypton; (5) the carbon dioxide; (6) the helium-neon and (7) ultra-violet. Lasers attached to microscopes are used in cytological and cytogenetic studies, research in embryology holography and in laser probe spectroscopy. In the field of biochemistry, the laser is available as a specialized high energy light instrument for all experiments in photochemistry in the synthesis of new compounds for basic studies in cellular morphology, including effects on DNA, RNA, enzyme systems, lysozymes, etc. Laser photography, especially holography, promises much for the future.

In the cancer research program, lasers are available for studies of (1) biochemistry with special interest in photobiology, dyes and chemotherapeutic agents; (2) tissue culture techniques; (3) animal cancer, induced, transplanted and spontaneous and (4) cancer patients, especially those with melanomas, multiple malignancies and accessible, so-called inoperative, cancers for which conventional treatment measures cannot be used.

In eye research, the laser retinal coagulator is available for (1) retinal tears and detachments; (2) angiomas; (3) retinoblastomas; (4) melanomas and (5) investigative studies especially in retinal vascular dynamics as concerns normal and pathologic states.

In dermatology, the laser is used for the treatment of (1) melanomas; (2) angiomas, especially the portwine type; (3) multiple skin tumors; (4) warts; (5) seborrheic keratoses or basosquamous acanthomas; (6) tattoos and (7) for investigative studies in radiation mechanisms optical properties of the skin, and protection.

In surgery, the laser is available for surgery of (1) melanomas; (2) inoperable malignancies; (3) blood vessel and heart surgery, and for bloodless surgery of such highly vascularized organs as the liver, spleen and lungs; (4) skin grafts and organ transplants; (5) as controls for plasma torch surgery and (6) for general surgical research.

In orthopedics, laser is available as an investigative tool for (1) bone drilling; (2) for accessible malignancies; and (3) as an additional instrument for investigative studies of the effects of radiation on the bone.

In the field of urology, the laser is available for investigative studies with (1) calculi and (2) as an adjunct of endoscopy.

In the field of gynecology, the laser is available for research on the cervix with colpomicroscopy. The goal here is to develop a laser operating colpomicroscope.

In maxillo-facial surgery, the laser can be used on the mastoid bone, in the middle ear, in the buccal cavity, on the larynx, and for head and neck tumors.

In neurosurgery for research in many areas and especially for thalamotomies.

In dentistry, the laser is used in preliminary studies for studies of the effect on enamel, caries and calculi. In addition, there are investigative studies on laser spectroscopy.

It is time now to expand well controlled studies of the laser effects on man. When effective safety control programs in well equipped special laser treatment centers are available, then the laser can be used.

The laser, then, has shown encouraging results in many diverse disciplines of the biomedical field. These encouraging results suggest that work should continue with due regard to proper safety programs and with instrumentation which is still cumbersome, expensive and sometimes erratic, but none the less fascinating. Future research, then, will determine the proper place of the laser in the physical sciences, in biology and in medicine.

Author and Name Index

226

Index

Subject Index